59 Topics in Current Chemistry
Fortschritte der chemischen Forschung

Organic Syntheses

Springer-Verlag
Berlin Heidelberg GmbH 1975

This series presents critical reviews of the present position and future trends in modern chemical research. It is addressed to all research and industrial chemists who wish to keep abreast of advances in their subject.

As a rule, contributions are specially commissioned. The editors and publishers will, however, always be pleased to receive suggestions and supplementary information. Papers are accepted for "Topics in Current Chemistry" in either German or English.

ISBN 978-3-662-15861-6   ISBN 978-3-540-37969-0 (eBook)
DOI 10.1007/978-3-540-37969-0

Library of Congress Cataloging in Publication Data. Main entry under title: Organic syntheses. (Topics in current chemistry; v. 59). Bibliography: p. Includes index. 1. Chemistry, Organic – Synthesis – Addresses, essays, lectures. I. Series.
QD1.F58 vol. 59[QD262] 540'.8s[547'.2] 75-25985

Originally published by Springer-Verlag Berlin Heidelberg New York in 1975
Softcover reprint of the hardcover 1st edition 1975

Contents

Springer-Verlag	D-6900 Heidelberg 1 · Postfach 105280 Telephone (06221) 487-1 · Telex 04-61723 D-1000 Berlin 33 · Heidelberger Platz 3 Telephone (030) 822001 · Telex 01-83319
Springer-Verlag New York Inc.	New York, NY 10010 · 175, Fifth Avenue Telephone 673-2660

Use of Activated Metals in Organic and Organometallic Synthesis

Dr. Reuben D. Rieke

University of North Carolina, Department of Chemistry, Chapel Hill, North Carolina 27514, U.S.A.

Contents

1. Introduction

The reaction of organic and inorganic substrates at a metal surface, either in a catalytic fashion or with the consumption of the metal, represents an extremely important area of chemistry. Ever since man's first discovery of this type of reaction, he has tried to increase the reactivity of the metal in order to allow the reactions to be carried out under milder conditions, to improve yields, or to extend the reaction to less reactive substrates.

In recent years, a number of new approaches to this old problem have appeared in the literature. We have recently developed a method for generating metal powders by the reduction of the metal salt with an alkali metal in an ethereal or hydrocarbon solvent.[1-8] The black metal powders generated by this procedure are frequently vastly superior to the commercial metal or standard activating techniques. Another approach which has received considerable attention is one that has evolved from Skell's carbon vapor reactor[9], namely the Metal Atom or Metal Vaporization Technique.[10] This approach utilizes the high chemical potential of metal atoms by co-condensing the metal vapor and the organic or inorganic substrate. The high chemical potential of the metal atoms coupled with the low reaction conditions have lead to many unusual new compounds. This "metal atom technique" will be discussed in greater detail later along with some of the older methods of activating metals. There will be no attempt to include the vast literature on the preparation of catalysts.

A major portion of this review will be concerned with the generation of activated metals using the reduction procedure developed in our laboratories. One goal of this article is to show how these new, highly reactive metals can be used to prepare new organic and organometallic compounds.

2. Generation of Activated Metals via the Reduction of Metal Salts

We recently discovered that the reduction of metal salts from ethereal or hydrocarbon solvents using alkali metals as reducing agents leads to highly reactive metal powders.[1-8] The reduction appears to be most conveniently carried out using an alkali metal and a solvent whose boiling point exceeds the melting point of the alkali metal. The metal salt to be reduced must also be partially soluble in the solvent and the reductions are carried out under an argon atmosphere. The reductions in most cases are exothermic, and in some cases very exothermic, and generally are complete within a few hours. Details for each specific metal will be discussed later. The reduction produces a fine black powder of the metal in question. In addition, one or more moles of an alkali salt are also produced during the reduction. Through the issue is far

$$MX_n + nK \rightarrow M + nKX$$

from settled, it would seem from present data that a composite polycrystalline material is being generated. The physical properties of the individual metals will be discussed in their respective sections. In addition to the above procedure, we have recently found that for certain metals the addition of simple alkali salts before the reduction is carried out yields balck metal powders of even higher reactivity.

The apparatus required is very simple. The reductions are usually carried out in a three-neck flask equipped with condenser, septum, heating mantle, magnetic stirring, and argon atmosphere. In some cases, the metal salt to be reduced is dried overnight in a flask under vacuum or else the anhydrous salt is added to the flask under argon. Freshly distilled, dry solvent is then added to the flask, generally with a syringe. Finally, the alkali metal is added and the resulting mixture heated to the melting point of the alkali metal. If the solvent is *tetrahydrofuran* (THF), this will be the boiling point of the solvent. However, for the higher boiling solvents, the reduction will probably be carried out below the boiling point of the solvents.

For most of the chemistry discussed below, *potassium* (Baker purified) was used as the reducing agent. Very impure potassium or sodium generally lead to much reduced reactivity of the activated metals. Use of sodium, of course, requires a higher boiling solvent. In most cases, the following reactions of the activated metals with organic and inorganic substrates were carried out in the vessel and solvent used for the reduction. In some cases, the solvent was stripped off and a different solvent added to carry out the reaction; these will be pointed out below. The alkali salts generated in the reduction process are not removed prior to any additional chemistry and in some cases it will be seen that they prove harmful to the desired reactions.

Even though we have had no trouble with fires or explosions caused by the activated metals, we would suggest *extreme caution* in working with these materials. We would suggest that until the person becomes very familiar with all the characteristics of the particular metal powder he is working with and the reduction reaction to generate the metal powder, extreme caution be taken at every step. We have found for every metal powder we have generated to date that, if they are removed from the reaction vessel wet with solvent, the metal powder will not spontaneously ignite. In spite of this, we would recommend that all metal powders be transferred under an argon atmosphere. If the metal powders are dried before being exposed to the air, some have been found to be pyrophoric such as magnesium (generated *via* the KI procedure discussed below) and aluminum. One potential problem can be the heat generated during the reduction process. This has only been observed in two cases where the metal salts in question were very soluble in the solvents used. Reduction of $ZnCl_2$ in THF[2] and also $FeCl_3$ in THF[11] were both fairly exothermic and extreme caution should be used at the start of the reduction of these and other very soluble salts. Use of sodium-potassium alloy as a reducing agent with these salts is not recommended and can lead to explosive conditions.

3. Activated Magnesium

3.1. General Procedure of Metal Salt Reduction Technique

The use of magnesium in organic, inorganic, and organometallic chemistry is very extensive. Grignard reagents represent one of the most important and versatile classes of organometallic intermediates known to the synthetic chemist. In spite of this, a vast number of Grignard reactions fail due to the unreactivity of the alkyl and aryl halide. Also, in some cases, coupling or eliminations plague the Grignard reaction.

Previous to our studies, there were three modifications of the general procedure for the direct synthesis of difficultly formed Grignard reagents from the reaction of magnesium metal and an organic halide: (1) use of higher reaction temperatures, (2) use of a more strongly coordinating solvent[12–16], and (3) activation of the magnesium metal.[17,21] The third method consists of activation of the magnesium by reduction of the size of the metal particle[20] or by a chemical reaction. The Gilman catalyst[17], which involves the addition of iodine to activate the magnesium, is representative of this technique. Ethylene bromide or ethyl bromide has been employed in catalytic amounts to activate the magnesium surface and in molar quantities as an entrainer.[18] Use of certain transition metal halides have proven to be useful catalysts.[19] Recently, Ashby combined the three techniques to prepare some alkylmagnesium fluorides.[21]

The conditions we have found most convenient for the preparation of the active magnesium involves the reduction of anhydrous $MgCl_2$ with potassium in refluxing THF under argon.[1,4] The reduction is exothermic and yields a black powder within minutes. The reduction seems to be complete after 30–45 minutes but reflex times of 1–3 hr are normally used. The black materials are exceedingly reactive and will react quantitatively with bromobenzene at –78 °C in five minutes. In fact, chlorobenzene will react at room temperature. The Grignard reactions are carried out by injecting the arylhalide or alkylhalide directly into the flask used to prepare the black materials. No attempt was made to remove the alkali salt generated in the reduction.

The reaction of the activated magnesium with *p*-bromotoluene[1] under a variety of conditions is shown in Table 1. The effects of reflux time, time at room temperature after reflux and prior to halide addition, and the ratio of magnesium to halide

Table 1. Reaction of *p*-bromotoluene with activated magnesium[1]

Reflux time, hr $MgCl_2 + K$[2]	Time at room temp, hr	Reaction temp, °C	Mg/*p*-BrC_6H_4-CH_3	% Yield Grignard, 5 min reactn time
1	0.5	–78[3]	1	87[4]
2.5	0.5	–78	1	86
3.5	0.5	–78	1	84
3.5	93	–78	1	73
91	0.5	–78	1	63
2.5	0.5	25	1	84
2.5	0.5	25	2	100
2.5	0.5	–78	2	100

[1]) All reactions run in THF.
[2]) Magnesium source for all reactions was $MgCl_2$ + K.
[3]) Dry Ice-acetone bath.
[4]) All reactions were complete in 5 min or less, so yields at longer times are not given. Yields were obtained by hydrolysis and vpc measurement of toluene formed.

were studied. As the reflux time employed in the reduction of $MgCl_2$ was increased, the yield of Grignard reagent decreased. Since little change in reactivity was noted for reflux times of 1–3.5 hr, subsequent reactions employed magnesium metal produced using reflux times of 2–3 hr. Also, an excess of $MgCl_2$ (10%) was usually run to prevent any problems with unreacted alkali metals. We have recently discovered that the optimum activity can be reached with an excess of $MgCl_2$ of 1–5%. The magnesium is also less reactive if it is prepared and then allowed to stand unstirred for periods longer than 0.5 hr. If the ratio of magnesium to (1) is one, the maximum yield is in the range of 84–87%. This ratio assumes complete reduction of $MgCl_2$ by K, and these results indicate that all of the magnesium produced may not be in the same state of reactivity. From our studies with other metals, it is likely that part of the potassium metal is being consumed in cleavage of THF. This fact plus the fact that the system was probably not totally free of moisture and oxygen would lead to a less than quantitative conversion of magnesium salt to magnesium metal. No attempt was made to find the exact proportions of potassium to magnesium salts and ultimately to the organic halide which would yield exactly 100% conversion of (1) to the Grignard. In order to obtain complete converion of the halide to Grignard, we simply used excess magnesium; in most cases we used a magnesium/(1) ratio of 2.

One point of particular interest was the conspicuous absence of coupling products. Free radicals are frequently suggested as intermediates in oxidative additions at metal surfaces and manifest themselves frequently in large yields of coupled products. Our fast reaction rates would lead one to speculate that large yields of coupled products might result. The complete absence of these products suggests that both the first and second electron transfer rates are extremely fast with this particular activated metal. This may prove to be particularly advantageous for those reactions which currently lead to nonstereospecific products due to the slow second electron transfer using ordinary magnesium.

3.2. Effect of Anion of the Metal Salt

In order to see what effect the anion had on the reactivity of the magnesium powder generated, several different magnesium salts, MgX_2, were reduced in K-THF.[4)] The resulting black powders were reacted with *p*-chlorotoluene at room temperature. For those salts which were very insoluble, the reduction failed and no black powders were generated; for example, MgF_2 and $MgSO_4$ did not work. Some of our results are listed in Table 2. Based on cost, availability, and slight differences in reactivity, we have found commercial anhydrous $MgCl_2$ to be most useful.

3.3. Effect of Solvent

$MgCl_2$ was reduced by K in the following solvents at their boiling point and the magnesium produced was then reacted with 1: THF, DME, benzene, Et_3N, and diglyme. These ease of reduction of $MgCl_2$ in these solvents was:

$$\text{THF} \simeq \text{diglyme} > \text{DME} > \text{Et}_3\text{N} > \text{benzene}.$$

Table 2. Reaction of *p*-chlorotoluene with activated magnesium prepared from various magnesium salts[1])

Magnesium Salt MgX_2	Reflux time, hr	Reaction temp, °C	% Yield Grignard at reaction time (min) of 5	10	30	60	120
$MgCl_2$	3	25	0	0	14	50	75
$MgBr_2$	4	25	0	0	21	47	
MgI_2	3	25	0	0	30	69	91

[1]) Mg/*p*-chlorotoluene = 2 reactions run in THF, using K.

The reactivity of the magnesium produced was:

$$\text{THF} > \text{DME} > \text{diglyme } (-78\ ^\circ\text{C}); \text{THF} \simeq \text{DME} > \text{diglyme } (0^\circ).$$

Comparisons were not made for Et_3N and benzene since $MgCl_2$ reduction was incomplete in these solvents. In addition to these solvents, other alkali metal-solvent systems also work, such as Na-diglyme, but little research with these systems has been done.

3.4. Effects of Added Alkali Metal Salts

In our studies to try to understand the effects of the alkali salt generated in the reduction on the reactivity of the metal powder, we added additional alkali metal salts either before or after the reduction of the $MgCl_2$. A very dramatic effect was observed when KI was added prior to the reduction of the $MgCl_2$. The reduction yielded a much deeper black material which was much more reactive than the magnesium powder generated in the initial process.[3,4]) This "KI-activated magnesium" powder was so reactive that it readily reacted with *p*-chlorotoluene at room temperature. Several different salts were tried to see if they had a similar activating effect. However, only in the cases of KI and NaI was any dramatic activation noted. Some of our results are listed in Table 3. For a more complete list of salts tried see Rieke and Bales.[4]) One additional experiment with I_2 as an activator was tried. When I_2 was added (Mg/I_2 = 20/1) after the reduction of $MgCl_2$, no Grignard was formed in 120 minutes at 0 °C. Thus it is apparent that the activation effect is from the presence of KI during the reduction process. The exact origin of this activation is currently under investigation in our laboratories.

3.5. Physical Properties of "Activated Magnesium"

All the factors which lead to the high reactivity of the "activated magnesium" or KI-activated magnesium are currently not understood. The deep black color indicates small particle size and hence large surface areas. The freshly prepared metal surface can be expected to be essentially free of an oxide coating. These factors alone would be expected to yield high reactivity. We have taken powder patterns of magnesium generated from $MgCl_2$-K-THF and find, interestingly, that only the characteristic pattern of KCl is observed.[22] This would indicate that either the magnesium is amorphous or that the magnesium particle size is 0.1 microns or less. We have taken scanning electron microscope photographs of both the "KI-activated magnesium", Fig. 1, and commercial 325 mesh magnesium, Fig. 2.[22] As can be seen,

Table 3. Effect of added salt, MX, on the reaction of *p*-chlorotoluene and activated magnesium[1])

MX	mg/mx	Reaction temp, °C	% Yield Grignard at reaction time (min) of 5	10	30	60	120
None		25	0	0	14	50	75
NaI	1	25	89	93	98	99	100
KI	1	25	87	93	98	99	100
KI[2])	1	25	0	5	42	71	85
KI[3])	1	25	43	66	92	97	100
KI	2	25	84	94	97	98	100
KI	20	25	51	76	94	97	100
KI	0.5	25	86	94	97	99	100
None		0	0	0	0	0	0
KI	10	0	8	17	55	75	85
KI	4	0	14	29	65	76	88
KI	2	0	26	46	73	85	91
KI	1	0	19	42	74	86	93
KI	0.5	0	25	45	74	85	93
LiI	2	0	0	0	0	0	11
NaI	2	0	0	6	21	49	70
RbI	2	0	15	27	65	79	89
CsI	2	0	6	12	41	73	89
I_2	20	0	8	10	40	54	64

1) Magnesium prepared from $MgCl_2$-K-THF, reflux 2–3 hr, Mg/*p*-chlorotoluene = 2, salt MX added prior to $McCl_2$ reduction.
2) MX added after $MgCl_2$ reduction.
3) Concentration of all reactants halved.

the overall size of particles in both samples is approximately the same. However, the "KI-activated magnesium" appears to be a conglomeration of small crystals. The scanning electron microscope does not differentiate between KI and Mg crystals. Therefore, until we can get some additional information such as X-ray fluorescence we can not single out the individual crystallites of KCl or Mg from the polycrystalline material. It is apparent, however, that due to the very rough nature of the particles very large surface areas are present.

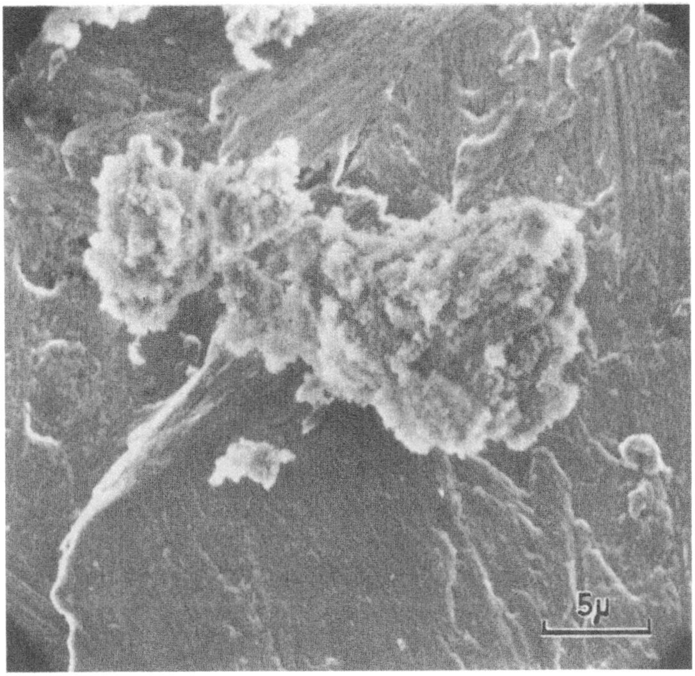

Fig. 1. Activated magnesium prepared by potassium reduction of $MgCl_2$ in the presence of KI

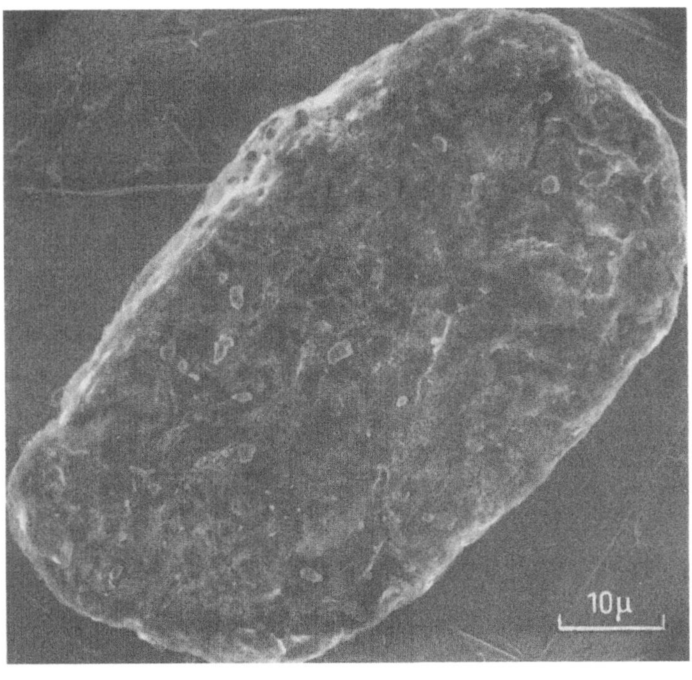

Fig. 2. Commercial magnesium (325 mesh)

3.6. Reactions of Activated Magnesium with Alkyl and Aryl Halides

We have examined the reactions of the "activated magnesium" and the "KI-activated magnesium" with various alkyl and aryl halides.[1-4] The results of these studies are summarized in Table 4.

Table 4. Reactions of activated magnesium with various halides[1]

Halide	Mg/ halide	Mg/KI	Reaction temp, °C	Reaction time, min	%Yield Grign.[2] Mono	Di	% Yield[3] $-CO_2H$
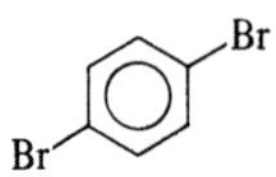	4	2	25	15		100	
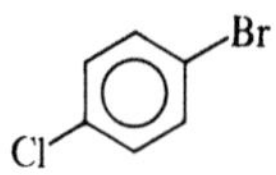	4	2	25	15	100	10	
				60	100	57	
				120	100	100	
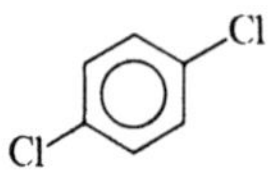	4	2	25	15	100	15	
				120	100	30	
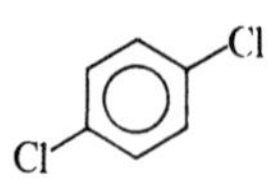	2	2	25	5	70	0	
				60	86	0	
				180	90	0	89[4]
	2		25	5	100		81[5]
	2		25	10	100		
				60			52[5]
	1.7	2	66	15	11		
				90	42		
				360	74		63[4]
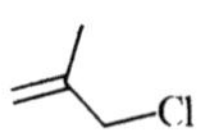	2		25	60			82[5]
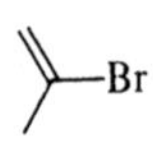	2		25	5	100		71[5]
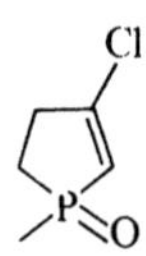	5		66	390			40

[1]) Mg source was either $MgCl_2$ + K or $MgCl_2$ + KI + K, THF solvent, refluxed 2–3 hr prior to halide addition.
[2]) Yield by vpc after hydrolysis.
[3]) Yield of isolated carboxylic acid.
[4]) Yield based on Grignard.
[5]) Yield based on starting halide.

Previous efforts to prepare dimagnesium derivatives of benzene have been successful only with dibromo- or bromoiodobenzene, required forcing conditions, and usually resulted in the monomagnesium derivative as the main product. This work has been surveyed by Ioffe and Nesmeyanov.[23] Using the $MgCl_2$-KI-K-THF system, we have prepared the di-Grignard of *p*-dibromobenzene in 100% yield in 15 min at room temperature. In earlier work, only one halogen atom of dichloro derivatives of benzene and naphthalene reacted with magnesium[24-27], and the chlorine of *p*-chlorobromobenzene (2) was found to be completely unreactive.[28] Using our method, we obtained a 100% yield of the mono-Grignard and a 10% yield of the di-Grignard of 2 in 15 min at room temperature. After 2 hr, the yield of di-Grignard was 100%. With *p*-dichlorobenzene (3) the yield of di-Grignard was 30% in 2 hr. Using a lower Mg/halide ratio permitted the rapid and selective formation of the mono-Grignard of 3, which was converted to 4-chlorobenzoic acid in 89% yield by CO_2 quench.

A primary halide, l-bromooctane, reacted as expected to give 100% yield of the Grignard in 5 min at room temperature. CO_2 quench gave an 81% yield of nonanoic acid. The tertiary halide 2-methyl-2-chloropropane gave a 100% yield in 10 min and was converted to pivalic acid in 52% yield after 1 hr of reaction. 1-Chlorobicyclo-[2.2.1]heptane (4) reacted slowly at room temperature, so this was rerun in refluxing THF. Again, the Grignard preparation was slow, giving a 74% yield after 6 hr of reflux. The Grignard was then quenched with CO_2 to give l-bicyclo[2.2.1]heptanecarboxylic acid (5) in 63% yield. Bixler and Niemann[29] prepared (5) from (4) by conversion of (4) to the lithium salt, followed by CO_2 quench. Our method appears to be superior since it obviates the preparation of lithium sand used in the procedure of Bixler and Niemann.

An allyl halide, 3-chloro-2-methylpropene (6), was converted to 3-methyl-3-butenoic acid (7) in 82% yield after 1 hr of reaction at room temperature. Wagner[30] converted (6) to the Grignard in 81% yield after 10 hr at 14–16°, and obtained (7) from the Grignard in 40% yield after CO_2 quench.

Finally, the vinyl bromide 2-bromopropene (8) was converted to the Grignard in 100% yield in only 5 min at room temperature using our normal magnesium system, $MgCl_2$-K-THF. The Grignard of (8) was then converted to methacrylic acid in 71% yield. Normant[31]), in his pioneering work on vinyl Grignard preparations, has recommended a 40–50° range for halide addition to magnesium in THF, followed by heating for 0.5–1 hr at 70–80°. By this procedure he was able to obtain yields of 95–97% in the preparation of vinylmagnesium bromide. Our activated magnesium appears to be an improvement of Normant's method, since it allows the preparation of vinyl Grignards at room temperature or lower.

Reaction of 1-methyl-3-chloro-2-phospholene-1-oxide (9) was particularly difficult even with "KI-activated magnesium". Reaction of 9 with „KI-activated magnesium" in refluxing THF for 6 1/2 hours gave approximately 40% of the corresponding carboxylic acid after treatment with carbon dioxide and then acid.

3.7. Reactions with Organic Fluorides

Recent work by Ashby[21] has shown that alkyl fluorides can, in fact, be converted to the corresponding Grignard reagents in high yields. He found that refluxing THF and magnesium activated with the Gilman catalyst (I_2) gave the best results. We have tried our "activated magnesium" and "KI-activated magnesium" with *p*-fluorotoluene (10) and 1-fluorohexane (11) under a variety of conditions. The results are summarized in Table 5. Perhaps this table best demonstrates the high reactivity of the "KI-activated magnesium". Prior to our work, all efforts to prepare the Grignard reagent from fluoro-benzene and magnesium had failed. Some of these attempts included reflux times of weeks or months. Using the $MgCl_2$-KI-K system, refluxing THF, and a Mg/10 ratio of four gave a 69% yield of the fluoro Grignard in 1 hr. Quenching the Grignard with CO_2 gave a 63% yield of the expected *p*-toluic acid. Using a lower Mg/10 ratio resulted in lower yields. Using diglyme but no KI gave only 8% yield after 1 hr at 162° (solvent refluxing). Continued refluxing in THF or diglyme for periods of 24–48 hr gave no appreciable increase in Grignard formation compared to reflux times of 2–3 hr.

Table 5. Reaction of *p*-fluorotoluene and 1-fluorohexane with activated magnesium[1,3]

Fluoride	Mg/fluoride	Mg/KI	Reaction temp, °C	Solvent	% Yield Grignard at reaction time (min) of			
					10	30	60	180
10[2]	1		162	Diglyme			8	5
10	2	2	66	THF		26	27	55
10	3	2	66	THF	40	53	59	58
10	4	2	66	THF	58	57	69	66
11	2		25	THF	10	26	27	54
11	2		66	THF	53	48	72	71
11	2	2	25	THF	38	53	73	89

[1]) Magnesium source for all reactions was $MgCl_2$ + K.
[2]) Refluxed 18 hr prior to fluoride addition.
[3]) All THF reactions refluxed 3 hr prior to fluoride.

Reaction with (11) using the normal procedure gave 54% yield in 3 hr at room temperature and 71% yield in 3 hr at 66°. When the KI procedure was employed, the yield was 89% after 3 hr at room temperature. Quenching the Grignard of (11) with CO_2 gave the expected heptanoic acid in 88% yield, based on Grignard. These results are superior to those obtained by Ashby[21]) for the preparation of the Grignard of (11), and provide further evidence for the high reactivity of magnesium prepared by our method.

3.8. Reactions of Grignard Reagents

Grignard reagents prepared with our activated magnesium appear to react normally with various substrates. Reaction with H_2O gives the hydrocarbon and with CO_2 the carboxylic acid.

Bromobenzene was converted to the Grignard with "activated magnesium", and the Grignard was then reacted with benzophenone (12) in analogy with the procedure of Bachmann and Hetzner.[32] Triphenylmethanol (13) was obtained in 87% yield, compared to the literature[32] value of 89–93%. Next, the Grignard was prepared at –78 °C (Dry Ice-acetone bath) and reacted with (12) at –78°. Reaction for 2.5 hr yielded (13) in 28% yield, and the yield was raised to 73% after 17 hr reaction at –78 °C.

3.9. Reaction of Activated Magnesium with Dienes

A recent report by Saloman[33] indicates that a mildly exothermic reaction occurs upon stirring a suspension of "activated magnesium" in THF under an atmosphere of butadiene. This mixture was then reacted with silicon tetrachloride to give 5-silaspiro-[4.4]nona-2,7-diene.

This is a vast improvement over the old procedure using butadiene, dichlorodimethylsilane, and ordinary magnesium powder in hexamethylphosphoric triamide and requiring several days at elevated temperatures.[34] The reaction must be carried out in an autoclave and gives a single annelation product. No mention was made in this paper if "KI-activated magnesium" had been tried.

3.10. Reaction with Ketones

Although we have not had ample time to carefully examine this reaction, all indications are that "activated magnesium" will be quite useful for the preparation of pinacols.[35] We reacted benzophenone with "activated magnesium" in THF at –78 °C for 18 hours. Work up gave 92% of the desired pinacol.

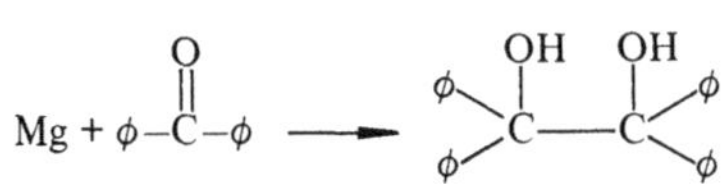

4. Activated Zinc

4.1. General Procedure for Preparation of Activated Zinc by Reduction Method

We have found that essentially the same procedure used to generate the activated magnesium can be used to generate "activated zinc". The most active zinc can be generated from anhydrous $ZnCl_2$, potassium, and THF.[2] The reduction is very exothermic and extreme care must be exercised during the initial stages. It is best not to stir the mixture during the first part of the reduction. The mixture should be heated very slowly at first and the heat should be removed just as the potassium melts. After the initial vigorous reaction is over, the reaction is heated at reflux with stirring for two hours. A deep black zinc powder is generated during the reduction. Particle size analysis indicates the average size is 17 microns. Powder patterns show both the characteristic lines of KCl and ordinary zinc metal.[36)]

A wide variety of zinc salts and various reducing agents have been tried but the above mentioned conditions seem to lead to the most active zinc. The addition of alkali salts prior to the reduction step does affect the activity of the zinc but this is still under investigation.

4.2. Reaction of "Activated Zinc" with Alkyl and Aryl Halides

The direct reaction of zinc metal with organic halides dates back to the work of Frankland[37–40)] He discovered that dialkyl zinc compounds could be prepared by the reaction of zinc metals with alkyl iodides. Several modifications have been suggested since that time to increase the reactivity of the metal. The majority of these modifications have employed zinc-copper couples.[41–45)] Other attempts to activate the metal have involve sodium zinc alloys.[48)] However, all of these procedures still only worked with alkyl iodides. Noller has used a mixture of alkyl iodides and bromides but found that the mixture must contain a large percent of alkyl iodide.[45)] Some recent work has, in fact, indicated that certain zinc-copper couples will react with alkyl bromides to give modest yields of dialkyl zinc compounds.[46,47)] However, all attempts to react zinc with aryl iodides or bromides have met with failure. The only way to make an aryl zinc halide or diaryl zinc compounds was by an indirect route, the two most common involving the addition of an aryl magnesium halide[49, 50)] or aryl lithium salt[51)] to a zinc salt.

We have found that the "activated zinc" will react rapidly with alkyl bromides in refluxing THF and, after distillation, give quantitative yields of the corresponding dialkyl zinc compounds. Prior to our work, there were no reports of the direct reaction of zinc metal with aryl halides. We have found that the "activated zinc" will react readily with iodobenzene and even with bromobenzene. The phenyl zinc halide does not disproportionate upon heating to give diphenyl zinc. It is felt that a stable ate-complex has been formed with the KCl present from the reduction. The phenyl zinc complex will, however, add to the carbonyl group of aldehydes to give the corresponding carbinol. Also, hydrolysis of the phenyl zinc complex produces benzene in high yields.

4.3. Reaction with α-Haloesters

Recent work in our laboratories had demonstrated that this zinc metal reacts very rapidly with α-bromoacetates at room temperature or even colder and the resulting organo zinc compounds react with carbonyl compounds (Reformatsky reaction[52)]) in high yields. The standard Reformatsky conditions generally employ benzene or benzene-ether solvents and are characterized by the difficulty of initiation of the reaction. Once the reaction has started there is considerable difficulty in moderating the exothermic reaction. The severe reaction conditions generally lead to only modest yields of 25–65%.[52,54] Two recent advancements in the Reformatsky reaction have involved the use of trimethylborate (TMB) – THF solvents[53] and the use of a zinc column.[54] The TMB-THF solvent does allow the reaction to be run at room temperature and the yields are good. However, TMB is not particularly cheap and the work-up is somewhat complicated by its presence. The zinc column method gives good yields but still requires refluxing benzene. We have found that the black zinc powders will react with α-bromoesters very rapidly (1/2 hr) at room temperature in diethylether. The yields of its addition to carbonyl compounds are in the range of 95% or better.[56]

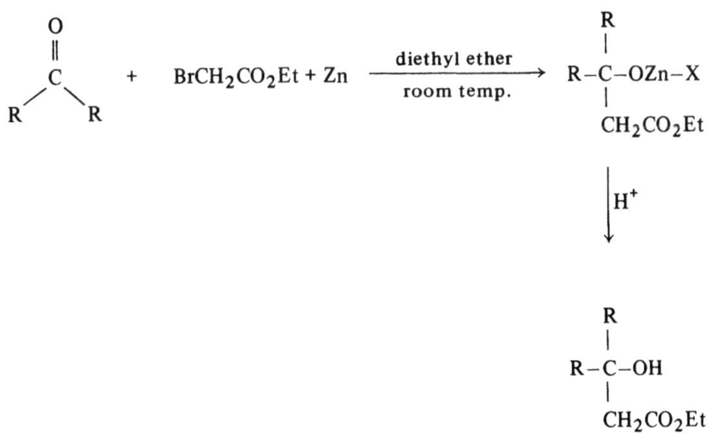

It is noteworthy that both aldehydes and ketones give high isolated yields of the resulting α-hydroxester. In this particular case, it was found that the reaction proceeded much better in diether ether than in THF. If THF is used to carry out the Reformatsky reaction with the activated zinc, yields on the order of 65% result. Thus the best results are obtained by first generating the activated zinc in THF using the standard proceedure. The THF is stripped off and dry diethyl ether is added. A one to one mixture of ketone or aldehyde and α-bromoester is then added dropwise at ice-bath temperatures. Finally, the reaction mixture is stirred at room temperature for one hour followed by normal work up procedures. Table 6 summarizes some representative reactions and also gives representative yields of the standard Reformatsky reaction, the zinc column method, and the TMB-THF method. The ability to use diethyl ether at room temperatures would appear to make the "activated zinc" procedure highly desirable.

The "activated zinc" will also react rapidly with α-chloroesters in THF. However, the yields are not nearly as high and generally are in the 65% range. In this particular case, the yields of the α-hydroxyester are very dependent on the anion of the zinc salt used to generate the zinc metal and can range down to as little as 5%.

Table 6. The yields of reformatsky reactions of some representative carbonyl compounds with the activated zinc and ethyl bromoacetate[1])

Carbonyl compounds	Conventional method	Zinc[54]) column	TMB-THF[53])	Activated zinc
1-Butyraldehyde	69[107])	(89)	90	97
1-Octylaldehyde	80[107])			78
Benzaldehyde	61[60])	(94)	95	98
Cyclopentanone	50[53])	(95)	87	97
Cyclohexanone	56[60])	(93)	80	95
p-Methylacetophenone	89[60])			98[2])

[1]) Glpc yields in parentheses, all others are isolated yields.
[2]) Isolated yield as β-hydroxy acid.

4.4. Zinc Couples

As mentioned in Section 4.2, zinc couples such as zinc-copper and zinc-silver couples have frequently been used in organic synthesis due to their increased reactivity. Their use in the preparation of dialkyl zinc compounds has already been pointed out. To try to summarize all the reactions in which the use of zinc couples has been reported, even for just the past few years, would be much to ambitious for this review. Therefore, we shall just present some representative examples and key papers reporting the latest advances in new types of couples.

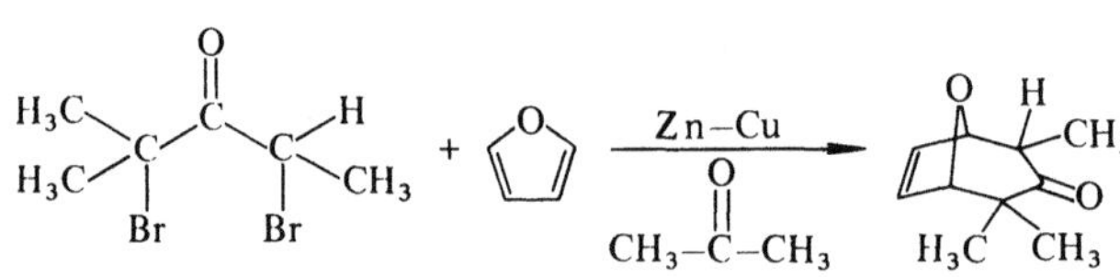

Ref.[98])

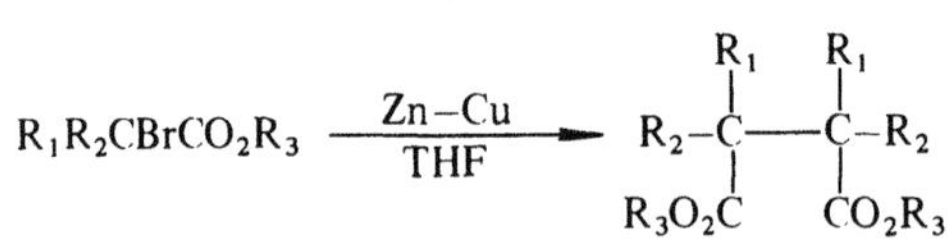

Ref.[99])

R_1	R_2	R_3	% yield
Me	H	Et	35
i-Pr	H	Me	40
Me	Me	Me	38

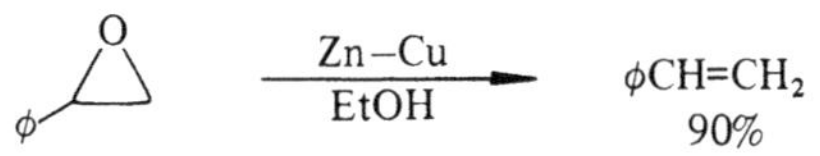

Ref.[100)]

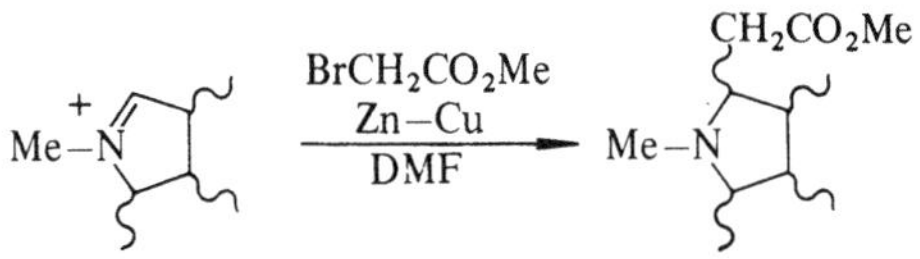

Ref.[101)]

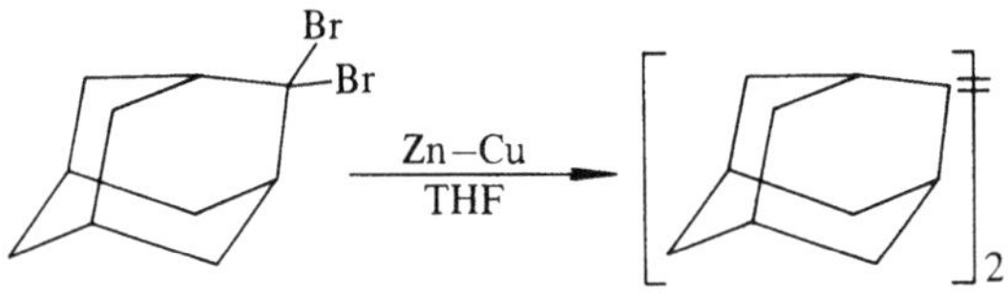

Ref.[102)]

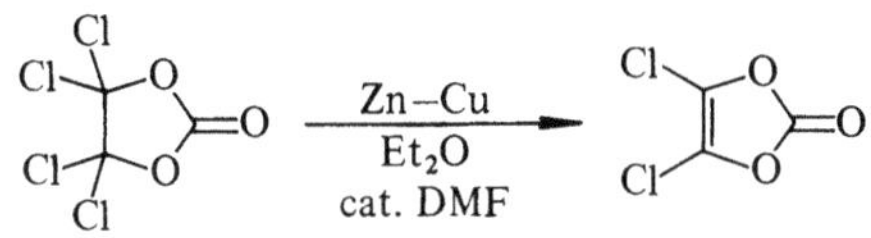

Ref.[103)]

A recent article by Swenton *et al.*[104)] reported deuterium incorporation via the reduction of halides with a zinc-copper couple. They found that freshly prepared zinc-copper couple in dry ether solvents with heavy water as the deuterium source effects clean reduction of dibromocyclopropanes to their monobromo derivatives. In a second system, they[104)] found that dichloroketene adducts underwent clean reduction to the corresponding cyclobutanones

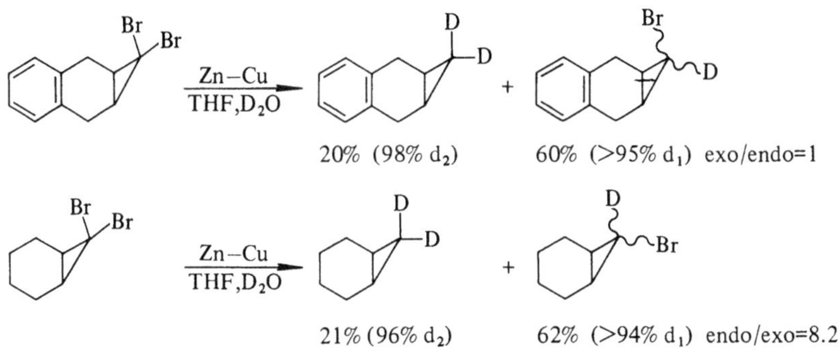

in good yield and with a high degree of deuterium incorporation. Thus, this method can serve as an alternative to tri-n-butyltin hydride[105)] or zinc-acetic acid[106)] reduction. Swenton *et al.* [104)] also reported that their freshly prepared zinc-copper couple could be used to reduce β-chlorovinyl ketones to the corresponding α, β-unsaturated ketone in high yields. One of the most prominent uses of zinc couples has been in the Simmons-Smith reaction[55)] shown below. The reaction, however, has often been plagued with poor reproducibility and low yields due mainly to difficulties in preparing the zinc-copper couple. We have found that the activated zinc prepared by reduction of zinc salts will in fact react with methylene iodide in

Table 7. Examples of improved Simmons-Smith reactions using zinc-silver couple and pyridine work-up

Olefin[1]	Reaction time, hr	Products	Yield[2]	The corresponding reactions using Zn/Cu couple[3]: Equivalents of CH_2I_2	Time hr	Yield	Ref.
 1	2	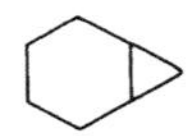 +1	95[4] 5	1.3 1.4	24 36	92 50	2)3) 3)
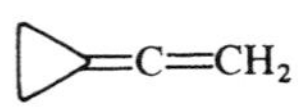 2	24	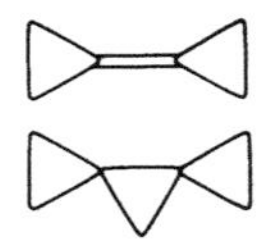 +2	36[4] 15[4] 47	6	24	38 5 57	9)
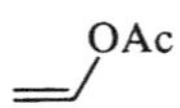 3	18	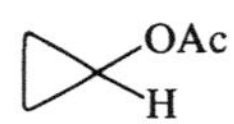 +3	30 70	0.5	20	31	1)
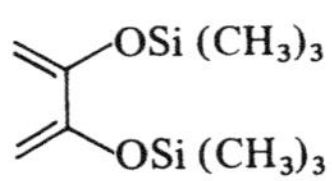 4	3	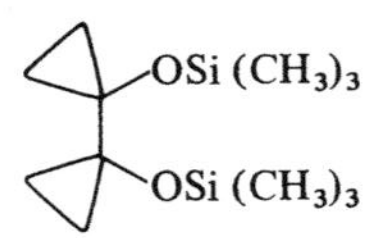	78	1.3	3	72[3]	

Olefin[1])	Reaction time, hr	Products	Yield[2]) %	The corresponding reactions using Zn/Cu couple[3]) Equivalents of CH_2I_2	Time hr	Yield %	Ref.
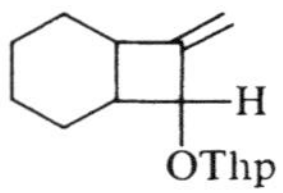 5	15	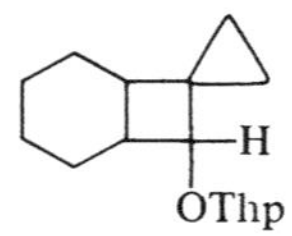	85				
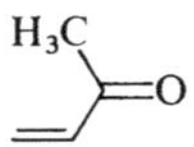 6	2	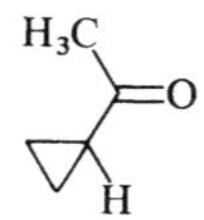	60	1.4	48	48	3)
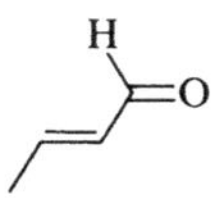 8	24	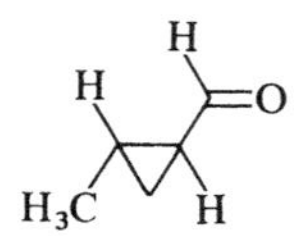 +8	15 85	1.4	20	0	3)2)

[1]) In both methods, 1.3 equivalents of diiodomethane were used for 1.0 equivalent of olefin, except in the case of allene. All reactions were conducted in boiling ether. The products were isolated by distillation.
[2]) The yields were determined by analytical and preparative G. L. C. of the distillate.
[3]) The reaction mixtures were worked up by acidic hydrolysis, except for the cyclopropanation of 4 and 8 (pyridine work-up).
[4]) The same yield is obtained when the cyclopropane derivative is extracted from the mixture by acidic hydrolysis.

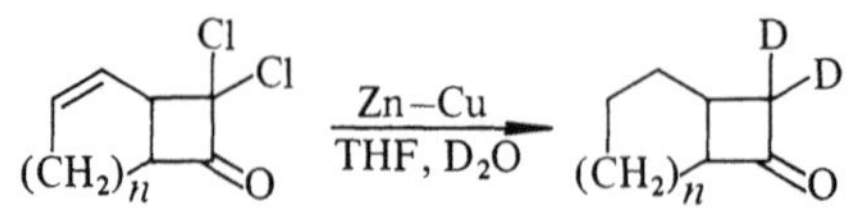

$n = 2$ $\quad$ $n = 2$, 80% (> 94% d_2)
$n = 1$ $\quad$ $n = 1$, 72% (> 95% d_2)

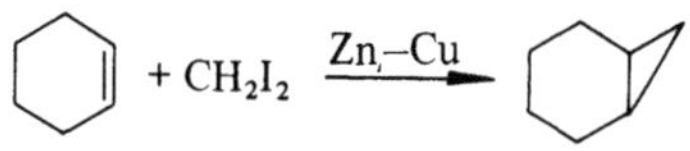

the presence of olefins to give Simmons-Smith product.[56] The yields, however, are only about 30% and thus are not very competitive with some of the recent advances reported for the Simmons-Smith Reaction. Rawson and Harrison[57] prepared a much improved zinc-copper couple by refluxing a mixture of zinc dust and cuprous chloride in ether. Olefins and methylene iodide were added directly to this mixture and yields of cyclopropanes of greater than ninety percent were reported. Denis, Girard, and Conia[58] have reported two modifications to the Simmons-Smith Reaction which have greatly extended the utility of this reaction, particularly with functionally substituted olefins. In place of a zinc-copper couple a zinc-silver couple is used and the usual final hydrolysis of the reaction mixture is replaced with the addition of an amine (*e.g.* pyridine) in order to remove zinc salts. Employing these modifications of the Simmons-Smith reaction, esters, aldehydes, ketones, and ether containing cyclopropane rings can be easily synthesized from the corresponding olefinic compounds. Some of their findings are given in Table 7. The preparation of the zinc-silver couple is described as follows. Granular zinc (17 g, 0.26 g-atom) is added to a stirred hot solution of silver acetate (100 mg) in acetic acid (100 ml). The mixture is stirred for 30 seconds and then the zinc-silver couple is isolated by decantation and washed with acetic acid and finally ether. Anhydrous ether is then poured onto the product and silver wool (2 or 3 small batches) is added. This zinc-silver couple appears to be more reactive toward diiodomethane than zinc-copper couples, gives better yields, and requires shorter reaction times.

Clark and Heathcock[59] have reported a recent modification of Conia's method and have found their zinc-silver couple to be very useful for the reduction of β-chloro-α, β-unsaturated ketones to α, β-unsaturated ketones.

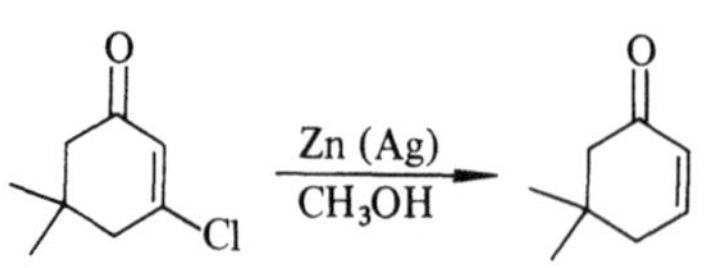

81% yield

In Heathcock's method zinc dust is stirred for 4 minutes with 10% aqueous HCl. The supernatant liquid is decanted and the zinc is washed with acetone and then ether. A suspension of anhydrous silver acetate in boiling acetic acid is then added. After the mixture is stirred for 1 minute, the supernatant is again decanted and the black zinc-silver couple is washed with acetic acid, ether and finally methanol. Some of the results they obtained are shown in Table 8.

Table 8. Reduction of β-haloenones

Reactant	Reaction time	Yield, [1]) %
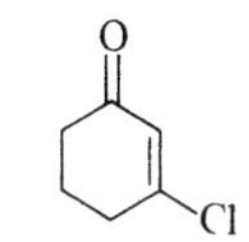	1.5 hr[2])	75
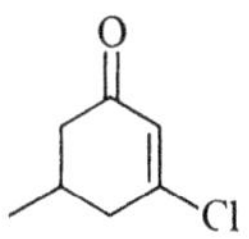	30 min	75
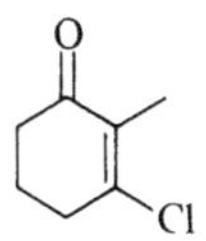	40 hr	77
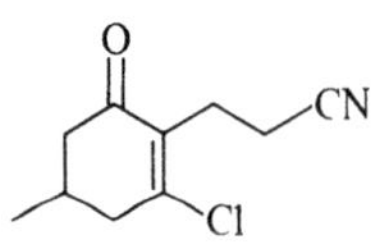	24 hr	93
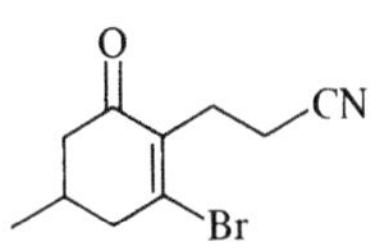	30 min	81
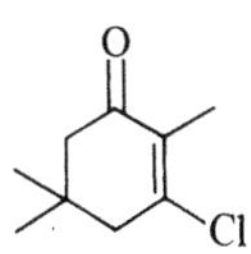	37 hr	[3])
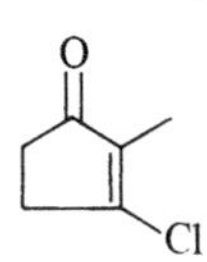	26 hr	65

[1]) Isolated yield.
[2]) Reaction carried out at 0°. At room temperature, the chloroenone reacts with methanol.
[3]) After 37 hr, glpc analysis shows that the original chloroenone is 58% reduced.

4.5. Activation of Commercial Zinc Dust

Any discussion of activated zinc should mention the classical method of activating commercial zinc dust. Shriner's conditions employed successively washing zinc dust with 20% HCl, water until neutral, then acetone, and finally anhydrous ether.[60] The resulting zinc dust was air dried and used immediately. Cornforth *et al.*[61] have described a modification of this procedure which involves the washing of zinc dust with 2% HCl, then ethanol, acetone, and finally ether. The resulting zinc dust is dried in vacuo with a crystal of iodine. They reported the use of this activated zinc in the Reformatsky reaction gave improved yields. Recently McIntosh and Beaumier[62] have reported using this activated zinc in the preparation of 1,3-cyclopentanedione in high yields.

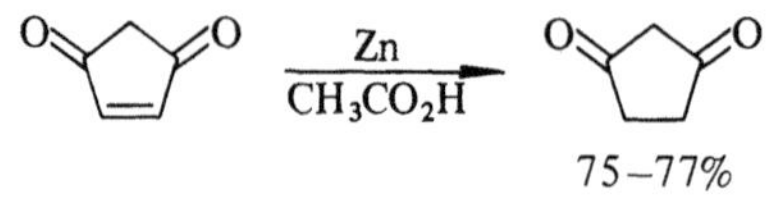

5. Cadmium

The reported use of cadmium metal in the literature has been very rare. However, the use of organocadmium compounds in organic synthesis is extensive. One of the primary reasons for their importance is their unique reactivity towards functional groups compared to other organometallic compounds. The usual manner of preparation involves reaction of an appropriate Grignard reagent or organolithium compound with a cadmium salt.[63,64] The preparation of organocadmium compounds by the direct reaction of cadmium metal and alkyl iodides has been reported;[65,66] however, this reaction has little synthetic value because of the long reaction times and low yields. A recent study has shown that alkyl iodides will react with metallic cadmium in HMPA to give good yields of solvated dialkylcadmium compounds.[67] Gaudeman[68] has recently reported an organocadmium analogue of the Reformatsky reagent from the reaction of *t*-butyl-α-bromoacetate with metallic cadmium in DMSO:

$$BrCH_2CO_2t\text{-}Bu + Cd \longrightarrow BrCdCH_2CO_2t\text{-}Bu$$

The organocadmium bromide was reacted with aldehyde in DMSO to give β-hydroxy esters but the yields were lower than the corresponding organozinc reagent.

6. Activated Indium

The use of indium metal in organic or organometallic synthesis is at best very limited. Schumb and Crane[69] reported that reaction of several alkyl halides with metallic indium was unsuccessful but that a slow reaction with methylene iodide did occur. It has also been reported that C_6F_5I will react with indium metal.[70] More recently Gynane, Waterworth and Warrall[71] reported the successful reaction of methyl, ethyl, and propyl bromides and iodides with metallic indium. The reaction, however, was very slow requiring 1–3 days for the iodides and 3–5 days for the bromides. Also, they found that the product was a mixture of R_2InX and $RInX_2$.

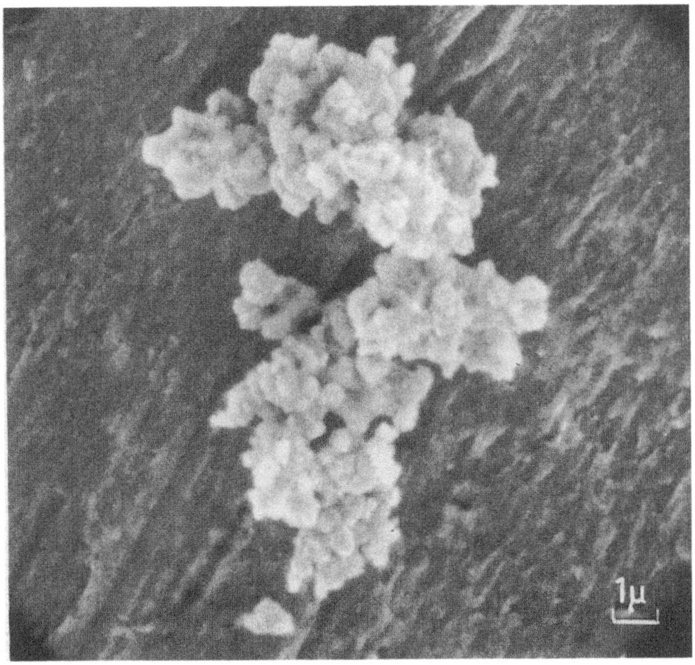

Fig. 3 Activated indium prepared by reduction of $InCl_3$ with potassium in xylene

We have found that the reduction of anhydrous $InCl_3$ with potassium in refluxing xylene produces a finely divided black indium metal of exceptional reactivity.[6,7] In this case ethereal solvents do not work as there appears to be extensive reductive cleavage of the solvent.

$$InCl_3 + 3M \longrightarrow In^* + 3MCl \quad M = K, Na$$

The reduction requires from 4 to 6 hours and is carried out under an Argon atmosphere. Particle size determinations on this material indicated a range of 10 μ to 0.2 μ with the average particle size being 4.0 μ. This is, of course, much smaller than that of commercially available indium powder (325 mesh). The activated indium shows a normal powder pattern of both indium metal and also the KCl. Scanning electromicroscope photographs of both the activated indium and commercial indium (atomized, 325 mesh) are shown below. Once again, the activated metal particles appear to be conglomerates of small crystals and have very high surface areas.

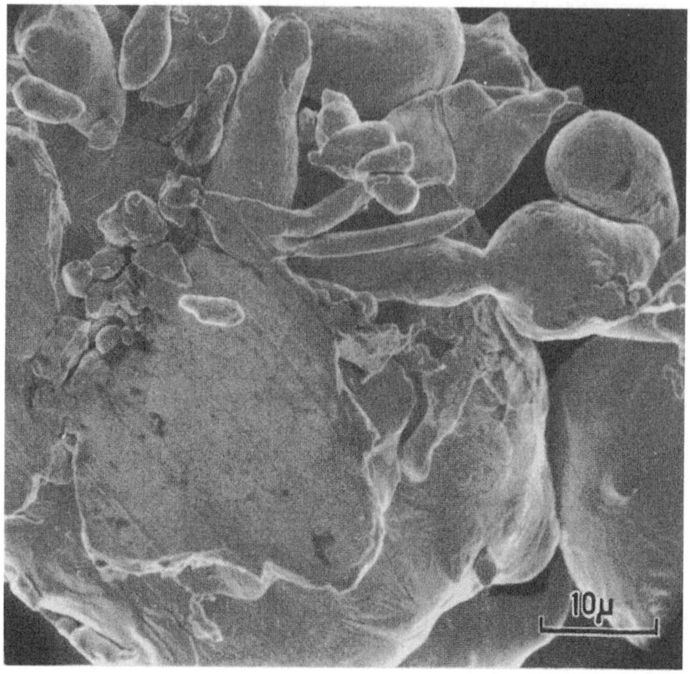

Fig. 4. Commercial atomized 325 mesh indium metal

Reaction of the activated indium with alkyl iodides was found to proceed very rapidly producing an almost quantitative yield of a single product, the dialkylindium iodide.[6] Furthermore, it was found that the

$$In^* + RI \xrightarrow[2\ hrs]{80\ °C} R_2InI + InI \qquad 97–100\%\ yield$$

R = Me, Et

activated indium would react with neat iodobenzene to produce high yields of diphenylindium iodide.[72] In fact, the activated indium will react

$$2In + 2\phi I \xrightarrow[150\ °C]{2\ hrs} \phi_2InI + InI \qquad 94\%$$

slowly with bromobenzene. Additional studies showed that the activated indium reacted readily with α-haloesters to give the organoindium product shown below.[73] This new Reformatsky reagent was found to add to the carbonyl group of ketones and aldehydes to give α-hydroxyesters in high yields.

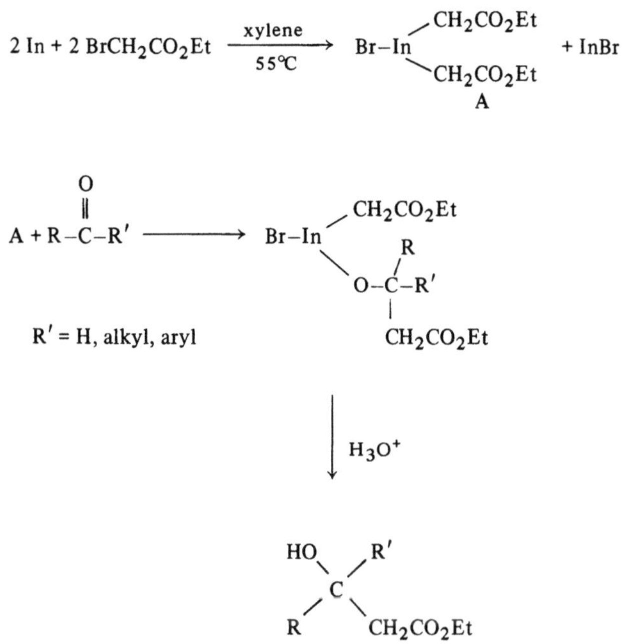

The reaction produces the best yields in nonpolar solvents and requires a one fold excess of ketone or aldehyde, most of which is recovered after the reaction. Also, the yield for alkyl aldehydes is poor, being on the order of 25%. The results for several ketones and aldehydes are summarized in Table 9. These results indicate the high reactivity of the indium powder generated by the reduction technique and it is expected that the above examples are only a beginning of the new reactions this metal powder will be useful in.

Table 9. Summary of reformatsky reaction of carbonyl compounds with activated indium and ethyl bromoacetate[1])

Carbonyl compound	Product	Solvent	Reaction temp. (°C)	% Yield[2])
Cyclohexanone	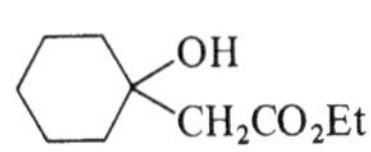	Xylene	55	100
Cyclopentanone	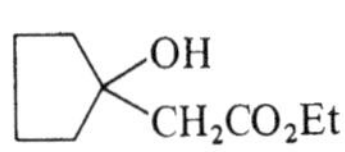	Xylene	55	80
Benzaldehyde	$C_6H_5CH(OH)CH_2CO_2Et$	Xylene	55	83–100
p-Methylacetophone	$CH_3C_6H_4C(OH)(CH_3)CH_2CO_2Et$	Et_2O	34	59

[1]) All reactions were carried out for 2 hrs.
[2]) Based upon glpc analysis using internal standard, $BrCH_2CO_2Et/2$ = 100%.

7. Activated Thallium

Despite much recent work on organic synthesis using thallium salts, there has been very little work using thallium(0). Aromatic nitro compounds are reduced to azoxy compounds by metallic thallium in refluxing ethanol.[74)] Diarylhalonium fluoroborates have been shown to react with metallic thallium to give diarylthallium fluoroborates; however, yields are low and reaction times long.[75)] Small yields of triethylthallium were obtained by direct reaction between ethyl chloride and finely divided alloys of thallium with sodium.[76)] Thallium has long been known to form thallous ethoxide when subjected to refluxing ethanol in the presence of oxygen over a long period of time.[77)]

Metallic thallium has been suggested as an intermediate in a reaction reported by Gilman and Jones.[78,79)]. The addition of an alkyl or aryllithium compound to a mixture of the corresponding alkyl or aryliodide and thallous halide give a dark precipitate of the finely divided metal. This disappeared during the course of the reaction to give good yields of the trialkyl or aryl thallium compounds. The overall stoichiometry of these reactions was as in Eq. (1).

$$2\,RLi + RI + TlX \rightarrow R_3Tl + LiI + LiX \qquad (1)$$

This was attributed to the sequence of steps in Eq. (2).

$$\begin{array}{lll} a) & 3\,RLi + 3\,TlX \longrightarrow & 3\,RTl + 3\,LiX \\ b) & 3\,RTl \longrightarrow & R_3Tl + 2\,Tl \\ c) & 2\,Tl + 2\,RI \longrightarrow & R_2TlI + TlI \\ d) & R_2TlI + RLi \longrightarrow & R_3Tl + LiI \end{array} \qquad (2)$$

TlCl was treated with two molar equivalents of methyl lithium, giving a finely divided black precipitate. The later addition of methyl iodide caused the precipitate to disappear, leaving a clear solution of trimethylthallium. Previous attempts to cause methyl iodide to react with ordinary forms of thallium (lump metallic, Na-Tl alloy, and Tl-Cu couple) had all failed.[80)] The above reaction worked also with iodobenzene, although the yield was lower (79%) and reflux at 55 °C was needed.

The only literature reference to just thallium and iodobenzene reacting together was in 1908.[81)] In ether there was no reaction at all. In refluxing iodobenzene (189 °C) a red coating (TlI in one of its two crystal forms) on the thallium was obtained after 20 hours. A very small yield of biphenyl was also obtained. Benzene was believed to be a major product but would have been volatilized under the reaction conditions. The reaction's slowness was attributed to the coarse state of subdivision of the thallium.

We have found that reduction of TlCl from xylene with potassium yields a finely divided black thallium powder of exceptional reactivity.[82)]

$$TlCl + K \xrightarrow[\text{24 hr}]{\text{refluxing xylene}} Tl + KCl$$

The choice of solvent and thallium salt seem to be particularly important. Reaction of this thallium powder with iodobenzene when the ratio of iodobenzene to thallium is 1.5 leads to high yields of biphenyl. The full potential of this coupling reaction has not been examined but it may prove

$$2\,\mathrm{Tl} + 2\,\phi\mathrm{I} \xrightarrow[140\,^{\circ}\mathrm{C},\ 24\ \mathrm{hr}]{\text{xylene}} \phi\text{–}\phi + 2\,\mathrm{TlI}$$

to be an attractive alternate to the Ullman reaction.

An interesting effect is noted if the reaction is run with an excess of thallium. In this case, with a ratio of iodobenzene to thallium of 0.3, the main product is triphenyl thallium as indicated by the final hydrolysis product.

$$4\,\mathrm{Tl} + 3\,\phi\mathrm{I} \longrightarrow \phi_3\mathrm{Tl} + 3\,\mathrm{TlI}$$

$$\phi_3\mathrm{Tl} \xrightarrow{\mathrm{HCl/H_2O}} \phi_2\mathrm{TlCl} \qquad \phi_3\mathrm{Tl} \xrightarrow{\mathrm{KI/H_2O}} \phi_2\mathrm{TlI}$$

The full potential of this highly reactive thallium metal has not been examined and it is expected that it will prove to be of value in many reactions.

8. Uranium, Thorium, and Plutonium

It has recently been reported[83] that di-π-cyclooctatetraene complexes of uranium, thorium, and plutonium can be prepared by the direct reaction of the finely divided metals with cyclooctatetraene. The finely divided metals were prepared by a procedure described by Seaborg and Katz.[84] Several cycles of formation and descomposition of the hydride gave a finely divided reactive metal. However, treatment of the pyrophoric uranium with cyclopentadiene under the same conditions used to form uranocene failed to produce any detectable amounts of the known compounds $U(C_5H_5)_3$ or $U(C_5H_5)_4$.

9. Other Metals

Highly reactive aluminum powders,[5] chromium powders,[8] and tin powders[85] have been produced by the reduction of metal salts in ethereal or hydrocarbon solvents. However, very limited use of these metals to date precludes any discussion of their usefulness in organic and organometallic synthesis.

10. Metal Vaporization

The advent of the metal vaporization technique[9] has allowed the ready production of many organometallic and inorganic materials which previously were unknown or which required very difficult, many-step procedures. Probably the most advantageous use of this technique has been in organotransition metal chemistry. As this area has recently been treated in an excellent review by Timms[10], we shall concentrate on those areas of most interest to organic synthesis. Developed initially by Skell[9,87], Timms[86], and Koerner von Gustorf[88], this technique involves the cocondensation of high temperature metal atoms with organic or inorganic substrates on the cooled walls (usually −196 °C) of the reaction vessel. The high temperature atoms are formed by vaporizing metals from resistively heated wires or crucibles[86], by arcing[9], by electron-gun heating[89], or by laser heating[88].

Skell and Girard[87] have reported the cocondensation of magnesium atoms and alkyl halides at −196 °C which, upon warming, yield unsolvated Grignard reagents with different properties than normal solvated Grignard reagents. For example, propyl Grignard reagent normally adds to acetone to give the tertiary alcohol. Skell's propylmagnesium compound, however, rapidly enolized acetone. Also, their methylmagnesium compound reacted with crotonaldehyde by 1,2-addition rather than 1,4-addition. Klabunde[90] has reported that the cocondensation of zinc atoms with fluorocarbons yields the nonsolvated fluoroorganic zinc compounds. Skell[91] also has reported the reaction of aluminum atoms with unsaturated alkenes to directly produce organoaluminum compounds. Recent reports by Skell[92,93] indicate that bisarene sandwich compounds of molybdenum and tungsten and also tris(butadiene) tungsten and molybdenum can be readily prepared by the metal atom technique. A number of papers have appeared concerning the reactions of Ni, Pt, and Pd[94–96] atoms with various alkyl and aryl halides.

Klabunde has recently shown that cocondensation of Mg[97] or Ni[96] with solvents such as THF or hexane yield, upon warming, active metal slurries. The Mg-THF slurries produced in this manner appear to be quite reactive and should prove to be of general value in the generation of Grignard reagents. Some typical results are shown in Table 10. Whereas the reactivity of this Mg is high, it is not nearly as reactive as the Mg produced by the potassium reduction of $MgCl_2$. The Mg-THF slurry is, however, free of alkali salts which in some cases may prove bothersome.

Table 10. Grignard reagent preparation using magnesium-THF slurries

Halide	Conditions	Time	Yield (%)
C_6H_5Br	THF, 25°	very rapid	
C_6H_5Br	Et_2O, 25°	very rapid	
C_6H_5Cl	THF, 25°	3 hr	85
C_6H_5F	THF, 66°	?	–
o-$C_6H_4Cl(CH_2CH{=}CH_2)$	THF, 25°	23 hr	80
C_6F_5Br	THF, −30°	45 min	77
$CH_2{=}CHCH_2Br$	THF, −40°	30 min	61
CF_3Br	THF, −30°	50 min	1

11. References

1) Rieke, R. D. and Hudnall, P. M.: J. Amer. Chem. Soc. *94,* 7178 (1972).
2) Rieke, R. D., Hudnall, P. M., and Uhm, S.: J. C. S. Chem. Comm. *(1973),* 269.
3) Rieke, R. D., and Bales, S. E.: "J. C. S. Chem. Comm. *(1973),* 789.
4) Rieke, R. D., and Bales, S. E.: J. Amer. Chem. Soc. *96,* 1775 (1974).
5) Rieke, R. D., and Chao, L.: Syn. React. Inorg. Metal-Org. Chem. *4,* 101 (1974).
6) Chao, L., and Rieke, R. D.: J. Organometal. Chem. *67,* C64 (1974).
7) Chao, L., and Rieke, R. D.: Syn. React. Inorg. Metal-Org. Chem. *4,* 373 (1974).
8) Rieke, R. D., Öfele, K., and Fischer, E. O.: J. Organometal. Chem. *76,* C19 (1974).
9) Skell, P. S., Wescott, Jr., L. D., Goldstein, J. P., and Engel, R. R.: J. Amer. Chem. Soc. *87,* 2829 (1965).
10) Timms, P. L.: Adv. Inorg. Chem. Radiochem. *14,* 121 (1972).
11) Rieke, R. D., and Page, G.: to be published.
12) Normant, H.: Compt. Rend. *240,* 1111 (1955).
13) Normant, H.: Bull. Soc. Chim. France *(1957),* 144.
14) Ramsden, H. E., Balint, A. E., Whitford, W. R., Walburn, J. J., and Serr, R. C.: J. Org. Chem. *22,* 1202 (1957).
15) Ramsden, H. E., Leebrick, J. R., Rosenberg, S. D., Miller, E. H., Walburn, J. J., Balint, A. E., and Serr, R. C.: J. Org. Chem. *22,* 1602 (1957).
16) Marvel, C. S., Wollford, R. G.: J. Org. Chem. *23,* 1658 (1958).
17) Gilman, H., and St. John, N. B.: Rec. Trav. Chim. Pays-Bas *49,* 717 (1930); Gilman, H., and Kirby, R. H.: ibid. *54,* 577 (1935).
18) Pearson, E, Cowan, D, and Becker, J. D.: J. Org. Chem. *24,* 504 (1959).
19) Respess, W. L., and Tamborski, C.: J. Organometal. Chem. *18,* 263 (1969).
20) Fuson, R. C., Hammann, W. C., and Jones, P. R.: J. Amer. Chem. Soc. *79,* 928 (1957).
21) Ashby, E. C., Yu, S. H., and Beach, R. G.: J. Amer. Chem. Soc. *92,* 433 (1970); Yu, S. H., and Ashby, E. C.: J. Org. Chem. *36,* 2123 (1971).
22) Rieke, R. D., and Chao, L.: to be published.
23) For a review, see Ioffe, S. T., and Nesmayanov, A. N.: Methods of elemento-organic chemistry, Vol. 2, p. 18. Amsterdam: North Holland Publishing Co. 1967.
24) Gomberg, M., and Cove, L. H.: Ber. *39,* 3274 (1906).
25) Pink, H. S.: J. Chem. Soc. *123,* 3418 (1923).
26) John, E., and John, N.: Rec. Trav. Chim. Pays-Bas *55,* 585 (1936).
27) Normant, H: C. R. Acad. Sci. *239,* 1510 (1954).
28) Krause, E, and Weinberg, K.: Ber. *62,* 2235 (1929).
29) Bixler, R. L., and Niemann, C.: J. Org. Chem. *23,* 742 (1958).
30) Wagner, R. B.: J. Amer. Chem. Soc. *71,* 3214 (1949).
31) Normant, H.: Advances in organic chemistry, Vol. II, pp. 37–38. New York: Interscience 1960.
32) Bachmann, W. E., and Hetzner, H. P.: Organic syntheses, Collect. Vol. III, p. 839. New York: Wiley 1955.
33) Salomon, R. G.: J. Org. Chem. *39,* 3602 (1974).
34) Dunogues, J., Colas, R., Dedier, J., and Pisciotti, F.: J. Organometal. Chem. *25,* 51 (1970).
35) Rieke, R. D., Bales, S., and Chao, L.: to be published.
36) Uhm, S. J.: Ph. D. thesis, University of North Carolina at Chapel Hill, 1975.
37) Frankland, E.: Ann. *71,* 171 (1849).
38) Frankland, E.: Ann. *99,* 333 (1856).
39) Frankland, E.: Ann. *71,* 213 (1849).
40) Frankland, E.: Ann. *85,* 329 (1853); Theyer, J.: J. Chem. Educ. *46,* 764 (1969).
41) Gladstone, J. H.: J. Chem. Soc. *59,* 290 (1891).
42) Job, A., and Reich, R.: Bull. Soc. Chim. France *33,* 1414 (1923).
43) Kurg, R. C., and Tang, R. J. C.: J. Amer. Chem. Soc. *76,* 2262 (1954).
44) Renshaw, R. R., and Greenlaw, C. E.: J. Amer. Chem. Soc. *42,* 1472 (1920).

45) Noller, C. R.: Org. Syn. *12*, 86 (1932).
46) Zakharkin, L. I., and Yu, O.: Iza. Akad. Nauk. SSSR, Otd. Khim. Nauk., *(1963)* 193.
47) Galiulina, R. F., Shabanova, N. N., and Petukhov, G. G.: Zh. Obshch. Khim. *36*, 1290 (1966).
48) See Methods of elemento-organic chemistry Vol. 3, p. 8. (ed. A. N. Nesmeyanov and K. A. Kocheshkov). Amsterdam: North Holland Publishing Co., 1967.
49) Gilman, H., and Brown, R. E.: J. Amer. Chem. Soc. *52*, 4480 (1930).
50) Sheverdina, N. I., Paleeva, I. E., Zaitseva, N. A., and Kocheshkov, K. S.: Dokl. Akad. Nauk. SSSR *155*, 623 (1964).
51) Wittig, G., Meyer, F. J., and Lange, G.: Ann. *571*, 167 (1951).
52) Reformatsky, S. N.: Zh. Russ. Fiz.-Khim. Obshch. *19*, 298 (1887).
53) Rathke, M. W., and Lindert, A.: J. Org. Chem. *35*, 3966 (1970).
54) Ruppert, J. F., and White, J. D.: J. Org. Chem. *39*, 269 (1974).
55) Simmons, H. E., and Smith, R. D.: J. Amer. Chem. Soc. *80*, 5323 (1958); *81*, 4256 (1959).
56) Rieke, R. D., and Uhm, S. J.: Synthesis *1975*, 452.
57) Rawson, R. J., and Harrison, I. T.: J. Org. Chem. *35*, 2057 (1970).
58) Denis, J., Girard, C., and Conia, J. M.: Synthesis *1972* 549.
59) Clark, R. D., and Heathcock, C. H.: J. Org. Chem. *38*, 3658 (1973).
60) Shriner, R. L.: Org. Reactions, *1*, 1 (1942).
61) Cornforth, J. W., Cornforth, R. H., Popjak, G., and Gore, I. Y.: Biochem. J. *69*, 146 (1958).
62) McIntosh, J. M., and Beaumier, P.: J. Org. Chem. *37*, 2905 (1972).
63) Krause, E.: Ber. *50*, 1813 (1971).
64) Wittig, G., Meyer, F. I., and Lange, G.: Ann. *571*, 167 (1951); Nesmeyanov, A. N., and Makarove, L.: Zh. Org. Khim. *7*, 2649 (1937).
65) Wanklyn, J. A.: J. Chem. Soc. *9*, 193 (1856).
66) Lohr, P.: Ann. *261*, 48 (1891).
67) Chenault, J., and Tatibouet, F.: C. R. Acad. Sci., Ser. C *264*, 213 (1967).
68) Gaudeman, M.: C. R. Acad. Sci., Ser. C *268*, 1439 (1969).
69) Schumb, W. C., and Crane, H. I.: J. Amer. Chem. Soc. *60*, 306 (1938).
70) Deacon, G. B., and Parrot, J. C.: Aust. J. Chem. *24*, 1771 (1971).
71) Gynane, M. J. S., Waterworth, L. G., and Worrall, I. J.: J. Organometal. Chem. *40*, 9 (1972).
72) Chao, L., and Rieke, R. D.: Syn. React. Inorg. Metal-Org. Chem. *5* (1975).
73) Chao, L., and Rieke, R. D.: J. Org. Chem. *40*, 2253 (1975).
74) McKillop, A., and Taylor, E. C.: Chem. Brit. *9*, 4 (1973).
75) Nesmeyanov, A. N., Tolstaya, T. P., and Isayeva, L. S.: Dokl. Akad. Nauk. SSSR *125*, 330 (1959).
76) Groll, H.: J. Amer. Chem. Soc. *52*, 2998 (1930).
77) Sidgwick, N. V., and Sutton, L. E.: J. Chem. Soc. *1930*, 1461.
78) Gilman, H. and Jones, R. G.: J. Amer. Chem. Soc. *68*, 517 (1946).
79) Gilman, H., and Jones, R. G.: J. Amer. Chem. Soc. *72*, 1760 (1950).
80) Rochow, E., and Dennis, L.: J. Amer. Chem. Soc. *57*, 486 (1935).
81) Spencer, J., and Wallace, M.: J. Chem. Soc. *93*, 1827 (1908).
82) Rieke, R. D., and Page, G.: to be published; Page, G.: Ph. D. thesis, University of North Carolina at Chapel Hill, 1974.
83) Starks, D. F., and Streitwieser, Jr., A.: J. Amer. Chem. Soc. *95*, 3424 (1973).
84) Seaborg, G. T., and Katz, J. J.: The actinide elements, p. 138 ff. New York: McGraw-Hill, 1954.
85) Rieke, R. D., and Uhm, S. J.: to be published.
86) Timms, P. L.: J. C. S. Chem. Comm. *1969*, 1033.
87) Williams-Smith, D. L., Wolf, L. R., Skell, P. S.: J. Amer. Chem. Soc. *94*, 4042 (1972); Skell, P. S., and Girard, J. E.: ibid. *94*, 5518 (1972).
88) von Gustorf, E. K., Jaenicke, O., and Polansky, O. E.: Angew. Chem. Internat. Edit. *11*, 532 (1972).
89) Benfield, F. W. S., Green, M. L. H., Ogden, J. S., and Young, D.: J. Chem. Soc., Chem. Commun. *1973*, 866.

90) Klabunde, K. J., Key, M. S., and Low, J. Y. F.: J. Amer. Chem. Soc. *94,* 999 (1972).
91) Skell, P. S., Wolf, L. R.: J. Amer. Chem. Soc. *94,* 7919 (1972).
92) Skell, P. S., Van Dam, E. M., and Silvon, M. P.: J. Amer. Chem. Soc. *96,* 626 (1974).
93) Silvon, M. P., Van Dam, E. M., and Skell, P. S.: J. Amer. Chem. Soc. *96,* 1945 (1974).
94) Klabunde, K. J., and Low, J. Y. F.: J. Organometal. Chem. *51,* 33 (1973).
95) Klabunde, K. J., Low, J. Y. F., and Efner, H. F.: J. Amer. Chem. Soc. *96,* 1984 (1974).
96) Klabunde, K. J.: Angew. Chemie, in press.
97) Klabunde, K. J., Efner, H. F., Satek, L., and Donley, W.: J. of Organometal. Chem. *71,* 309 (1974).
98) Hoffman, H. M. R., Clemens, K. E., and Smithers, R. H.: J. Amer. Chem. Soc. *94,* 3940 (1972).
99) Fouquey, C. and Jacques, J.: Synthesis *1971,* 306.
100) Kupchan, S. M., and Maruyana, M.: J. Org. Chem. *36,* 1187 (1971).
101) Granelli, I., and Leander, K.: Acta Chem. Scand. *24,* 1108 (1970).
102) Geluk, H.: Synthesis *1,* 652 (1970).
103) Schart, H., Pinake, W., Feilen, M., and Droste, W.: Chem. Ber. *105,* 554 (1972).
104) Blankenship, R. M., Burdett, K. A., and Swenton, J. S.: J. Org. Chem. *39,* 2300 (1974).
105) Ghosez, L., Montaigne, R., Roussel, A., Vanlierde, H., and Mollet, P.: Tetrahedron *27,* 615 (1971).
106) Reyn, M., Huber, U. A., and Dreiding, A. S.: Tetrahedron Lett. *1968,* 3583.
107) Frankenfeld, J. W., and Werner, J. J.: J. Otg. Chem. *34,* 3689 (1969).

Received January 28, 1975

New Directions in Aromatic Nucleophilic Substitution

Prof. Dr. John A. Zoltewicz

Department of Chemistry, University of Florida, Gainesville, Florida 32611, U.S.A.

Contents

1. Introduction

Recent developments dealing with nucleophilic substitution at the annular carbon atoms of carbocyclic and heterocyclic aromatic molecules are examined. Emphasized are areas which the author believes will command the attention of more and more chemists and therefore will grow in importance. They are perceived to be new directions. Fascinating problems dealing with reaction mechanisms and useful synthetic methods are highlighted. Selected developments largely during the period 1972–1974 are considered; little earlier information will be presented because reviews[1, 2] are available. The subject is restricted to carbocyclic molecules containing six-membered rings and to heterocyclic compounds with five- and six-membered rings.

Subject matter is organized along two broad lines according to the timing of bond making and breaking. Considered first are reactions in which the new bond forms before the old one breaks; then follow transformations in which cleavage preceeds bond formation. Ionic and radical ionic reactions are included.

2. Intermediate σ-Complex Mechanism: $S_N Ar$

One of the oldest and most useful methods of substitution at aromatic carbon involves the $S_N Ar$ route in which a nucleophile first adds to a ring to give a σ-complex as an intermediate. Product forms when the leaving group, termed a nucleofuge,[3] is eliminated with its bonding pair of electrons.[1,2] Commonly employed nucleofugic leaving groups include

$$F, Cl, Br, I, RSO_3, RSO_2, RSO, RS, R_2S^+, R_3N^+, R_2N, RO \text{ and } NO_2.$$

The nature of the leaving group, the nucleophile[4–7] and the solvent[8, 9] can have an important influence on substrate reactivity. Generalizations are too lengthy to be included here; discussions are available.[1, 2, 10]

In order to have substitution proceed at a convenient rate under mild conditions the presence of one or more electron withdrawing substituents or heteroatoms at annular sites often is required. Such groups or atoms stabilize the intermediate complex and associated transition states. Illustrative are reactions between 1-fluoro-2,4-dinitrobenzene and amines, termed[11] aminodefluorinations, Eq. (1). Such mild

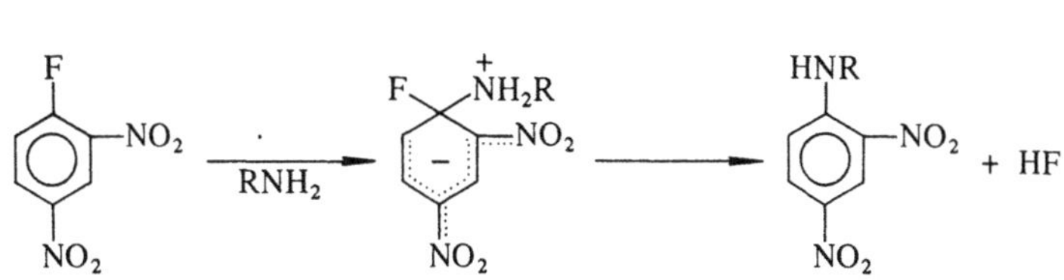

(1)

conditions are required that the reaction is useful in protein sequencing studies in which the free amino end of a protein or a peptide is converted to a dinitrophenyl derivative.[12] The two nitro groups provide considerable stabilization by a resonance effect of the negative charge and make substitution facile.

By comparison, the reaction of 3-chloro- or 3-bromopyridine with methoxide ion requires more severe conditions even though methoxide ion is a better nucleophile than an amine. The annular nitrogen atom provides only modest inductive stabilization and so temperatures on the order of 218 °C are required to effect substitution at convenient rates. Under these conditions the 3-methoxypyridine product is unstable, undergoing further reaction with methoxide ion at the side chain to cleave

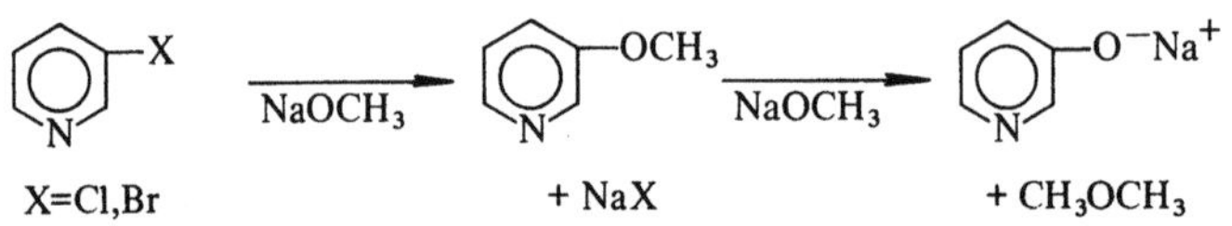

(2)

the ether linkage, Eq. (2). This substitution reaction therefore does not constitute a viable preparation of 3-methoxypyridine but it may be used to prepare the ether cleavage product 3-hydroxypyridine.[13] When aqueous hydroxide is employed with the halopyridines the temperature must be higher, about 300 °C in order to effect rapid substitution.[14]

Sometimes it is possible to detect a σ-complex formed during substitution. Using a stopped-flow technique it is possible to detect the transient formation of *1* which is the conjugate base of the σ-complex resulting from addition of *n*-butylamine to 2,4-dinitro-1-naphthyl ethyl ether in dimethylsulfoxide (DMSO).[15]

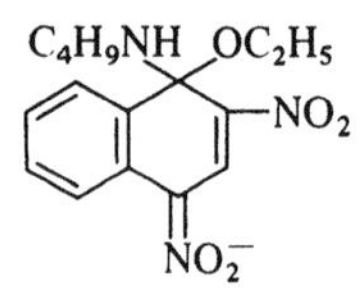

1

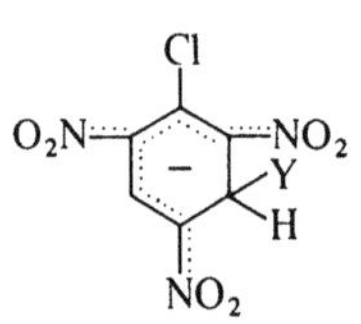

2

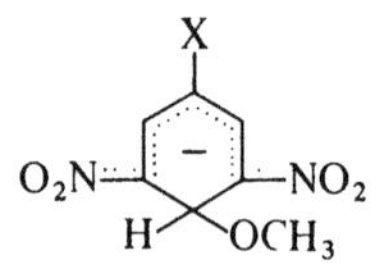

3

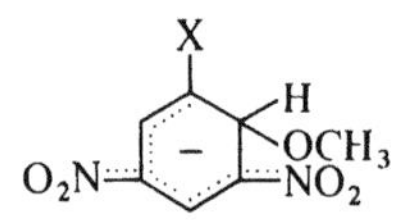

4

Two σ-complexes may form during a substitution reaction; the first, being incapable of undergoing substitution, reverts back to reactants which form a new complex. Thus, picryl chloride first reacts reversibly with a variety of nucleophiles in DMSO at unsubstituted carbon atom 3 to give complex *2* reflecting kinetic control. Substitution takes place by a slower irreversible reaction when a nucleophile adds to position 1 which contains the chloride ion leaving group.[16]

Anionic σ-complex formation has been studied with substrates which do not give rise to substitution products. Kinetically and thermodynamically controlled σ-complexes often are observed, NMR serving as a favorite method of detection and structure determination. 1-Substituted 3,5-dinitrobenzenes in methanol first add methoxide ion at position 4 to give *3*; this kinetic product then reverts to thermodynamically more stable adduct *4* having methoxide ion bonded to position 2. Values for the association equilibrium constants for the more stable adducts are highly dependent on the identity of the substituent and vary from $1 \cdot 10^{-6}$ to $8.9 \cdot 10^{-4}$ to 1.0 to 17 l/mol at 25° when X = H, I, CN and NO_2, respectively.[17]

Addition of amide ion to azines in ammonia gives anionic σ-complexes *5*–*7*[18] while similar reactions in the case of diazines yield *8*–*10*.[19] Complex formation in

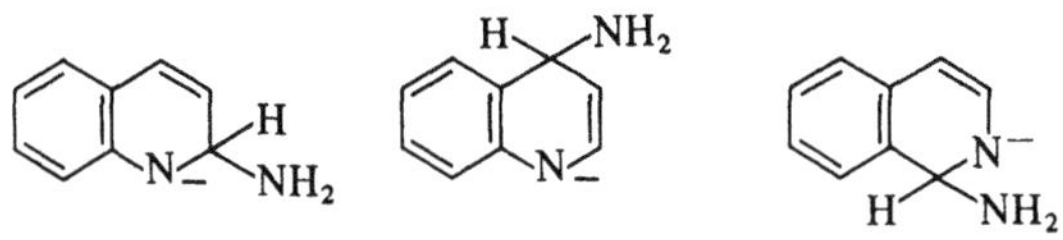

5 *6* *7*

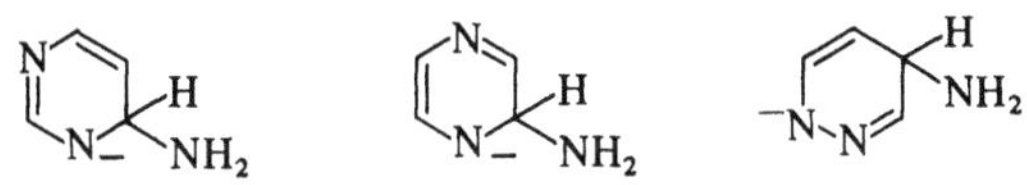

8 *9* *10*

all these cases is rapid and quantitative. With quinoline both kinetic and thermodynamic adducts are observed, addition of amide ion to position 2 to give *5* being faster than reaction at position 4 to yield the more stable complex *6*. It is worth noting that acidic substituents, including alkyl groups, when bonded to an aromatic ring may inhibit or prevent such complex formation by ionizing.[20, 21]

Approaches such as those considered in the following sections are being developed in order to circumvent the requirement that activating groups must be bonded to an aromatic ring to facilitate substitution.

2.1. Crown Ethers

Crown ethers, a class of macrocyclic polyethers,[22, 23] have the remarkable ability to solubilize alkali metal salts in solvents of low polarity. The crown ether solvates a metal cation by binding the ion to its oxygen atoms by means of ion-dipole interactions. The metal ion becomes located near or inside the cavity of the ether; illus-

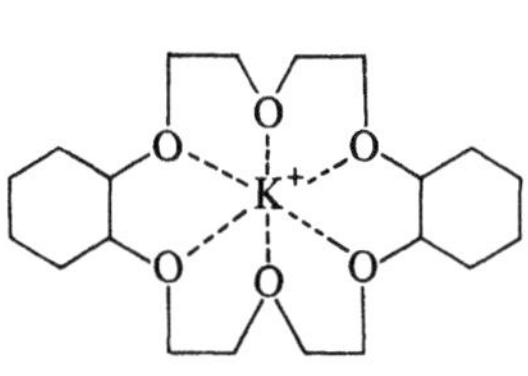

11

trative is *11*, a complex between potassium ion and dicyclohexyl-18-crown-6. By contrast, the anion of the salt is only poorly solvated ("naked"[24]) and consequently is unusually reactive.

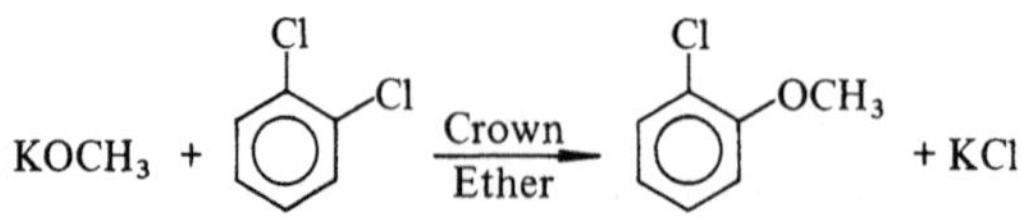

(3)

A crown ether can facilitate substitution between an unactivated aryl chloride and potassium methoxide. Thus, heating a solution of the dicyclohexyl-18-crown-6 complex of potassium methoxide in *o*-dichlorobenzene at 90° gives rise to 40–50% of *o*-chloroanisole, Eq. (3). No reaction is observed in the absence of the ether.[25] This approach is not always successful, however. The same ether failed to induce a reaction between potassium cyanide and *o*-dichlorobenzene in acetonitrile.[24]

Crown ethers can influence the state of aggregation of a reagent and thereby change its reactivity. A rapid reaction between *p*-fluoronitrobenzene and equimolar amounts of potassium *tert*-butoxide and 18-crown-6 ether in *tert*-butyl alcohol takes

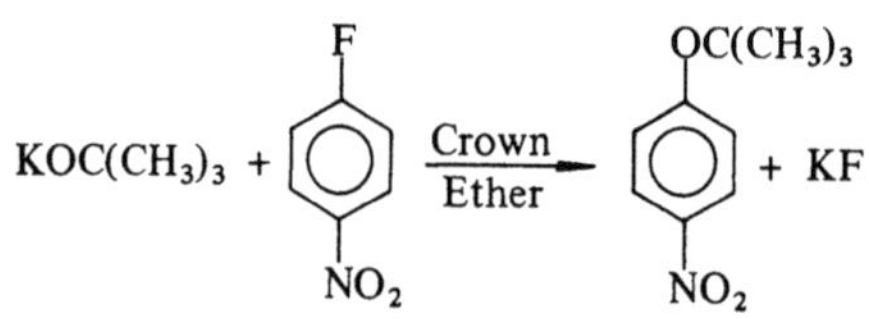

(4)

place to yield *p*-*tert*-butoxynitrobenzene, Eq. (4). But in the absence of the ether the rate is decreased 2000-fold. The ether changes potassium *tert*-butoxide from a tight ion pair to a loose, ether-separated pair which is more nucleophilic. Moreover, the ether has a negligible influence on the rate of reaction between the same substrate and potassium methoxide in methanol where ion pairing is unimportant.[26]

2.2. Polar Aprotic Solvents

Accelerated rates of substitution often result when reactions are carried out in polar aprotic rather than in hydroxylic solvents. This largely is a consequence of poor solvation of centers of negative charge in the aprotic media. By contrast, in hydroxylic media solvation of such centers by hydrogen-bonding is important.[27] Commonly employed aprotic solvents include DMSO, N,N-dimethylformamide (DMF) and hexamethylphosphoric triamide (HMPA). Thus, the reaction of 2-bromopyridine with methoxide ion at 110 °C to give 2-methoxypyridine is 3590 times faster in HMPA than in methanol.[28]

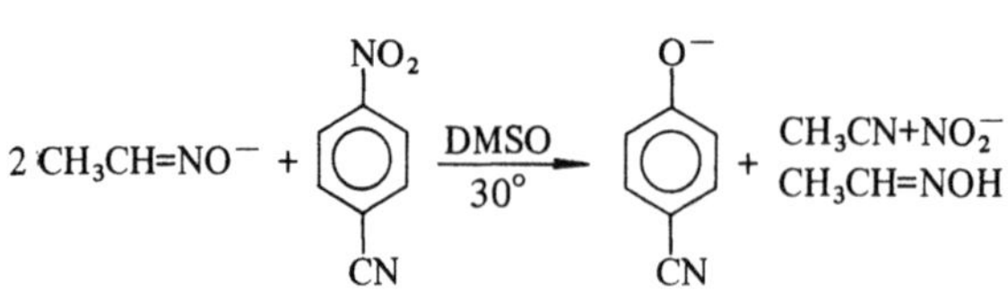

(5)

Displacement of a nitro group from a benzene ring bearing only a single activating group has not been common. However, using DMSO and two equivalents of the anion of acetaldoxime a rapid substitution of *p*-cyanonitrobenzene takes place to give *p*-cyanophenol, Eq. (5). The oximate ion displaces nitrite ion in the first step and the resultant intermediate undergoes an elimination reaction with a second oximate ion to give phenoxide ion product and acetonitrile. When ethanol is the solvent, no reaction occurs, even on heating. A direct approach to the phenol from the nitro compound and sodium hydroxide in DMSO fails.[29]

Naphthyl ethers[30–32] and sulfides[33] are easily prepared from fluoro compounds and the appropriate nucleophile in DMSO. Especially attractive is the synthesis of ethers with a *tert*-butyl group because the usual Williamson synthesis fails. 1-Fluoro- and 2-fluoronaphthalenes when heated with potassium *tert*-butoxide in DMSO yield their respective ethers along with ether cleavage products naphthol and isobutylene. Similarly, 3- and 4-fluoropyridines afford 3- and 4-*tert*-butoxypyridines, respectively, in about 60% yield.[34]

When carrying out substitution reactions with halides in polar aprotic solvents it is well to keep in mind some possible complications which include reductive dehalogenation and halogen migration. For example, 2,5-dibromothiophene (*12*) reacts exothermically with sodium methoxide in DMSO to give 2-bromothiophene (*13*) in

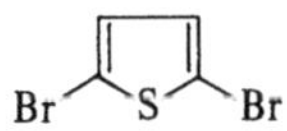

12

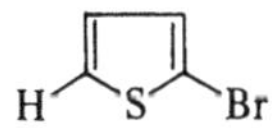

13

about 70% yield.[35] The mechanism probably is the same as that for aryl halides. This involves attack of ionized DMSO on halogen to give a carbanion which then is protonated.[36] The same dibromothiophene reacts with sodium methoxide in DMF to yield a mixture of mono-, di- and tribrominated thiophenes. In this case transfer of a bromine atom from ring to ring takes place when a hetaryl anion, formed by the action of methoxide ion, attacks a bromine atom bonded to another ring.[35]

2.3. Phase Transfer Catalysis

A development showing considerable promise for synthetic chemistry makes use of phase transfer catalysis.[37, 38] Reactions are carried out in a two phase system consisting of concentrated aqueous alkali and an organic layer which may be a neat reactant. The success of the method lies in the use of a catalyst, often a tetralkyl ammonium ion, which transfers hydroxide ion from the aqueous phase into the organic phase where it reacts with the substrate to generate ionic nucleophile. In the organic phase the nucleophile, paired with the tetralkyl ammonium ion, is especially reactive because it is poorly solvated. Although the nucleophile may be present in the organic phase in low concentration rapid transfer of the catalyst between the two phases gives rise to rapid reactions. The method also has considerable potential for industrial syntheses because it is simple to employ and does not require expensive solvents.

The first comprehensive study applying phase transfer catalysis to aromatic nucleophilic substitution appeared in 1974 and deals with carbanion nucleophiles derived from phenylalkanenitriles. The catalyst is triethylbenzylammonium chloride (TEBA).[39]

A series of chloronitrobenzenes reacts with the anionic form of phenylalkanenitriles to displace chloride ion. Although *para* chloro-nitro compounds are indicated in Eq. (6), *ortho* analogs react similarly. Electron donating or withdrawing groups may be bonded to the haloaromatic ring. Conditions are mild. Mixtures are kept at 40–50° for 3–4 hours. Yields of substitution products rarely fall below 60%; yields in excess of 90% are not uncommon. With dichloro compounds the chlorine atom *para* to a nitro group is more reactive than one at an *ortho* position.

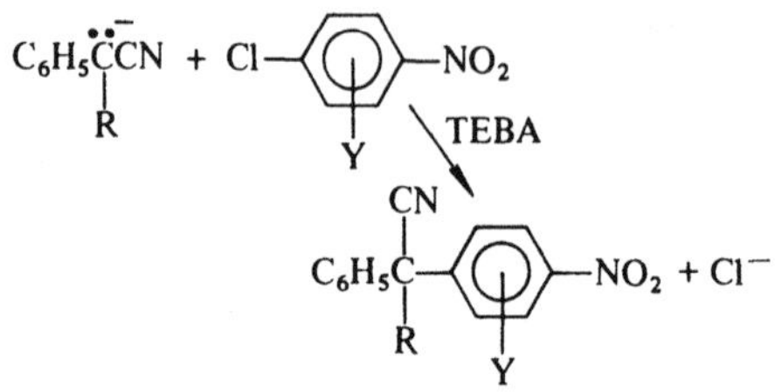

(6)

4-Halo-4′-nitrobenzophenones preferentially react with the conjugate bases of phenylalkanenitriles at the nitro group to give substitution products, Eq. (7). Nitro group displacement is general for 4-substituted 4′-nitrobenzophenones except when the nitrile contains a *sec*-alkyl group, electron transfer to give azoxy compounds being favored over substitution. Reaction conditions for the benzophenones are similar to those for the chloronitrobenzenes. Again yields of substitution product are high, being over 60%.

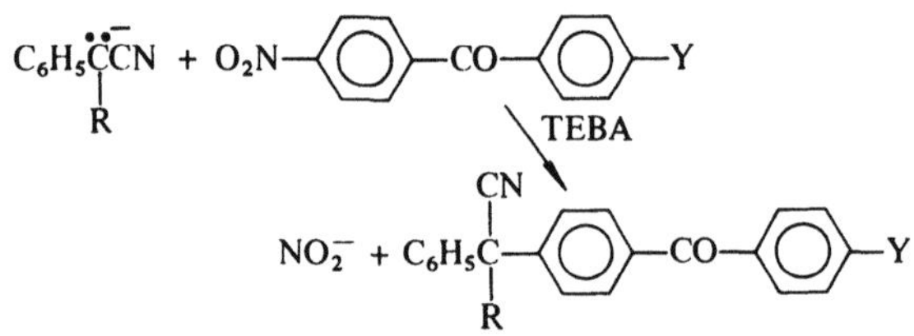

(7)

For comparison purposes some substitution reactions were tried under different conditions. While the use of DMSO and sodium hydride or alkali metal hydroxides allowed substitution products to be formed in yields comparable to those employing TEBA, other solvent-base systems failed. These include amide ion in ammonia and sodium hydride in tetrahydrofuran. If any substitution product were isolated using these other methods, the desired product was accompanied by much side-product. Perhaps the reason for successful substitution by methods employing phase transfer catalysis and DMSO solvent lies in the state of aggregation of the carbanion. Free ions and ion pairs are present under the successful conditions but higher aggregates are likely when failure results.[39)]

Another catalyst may be employed in place of TEBA; this is a crown ether such as dibenzo-18-crown-6. Addition of only 1 mole percent to a two phase system promoted reactions such as those given by Eqs. (6) and (7).[40)]

2.4. Activation by Chromium Tricarbonyl

Because nucleophilic substitution is facilitated by the presence of electron-withdrawing groups, it is desirable to be able to add an activating group and then to remove it following substitution. Introduction and removal of such substituents is not always easy. An attractive route involves the use of a chromium tricarbonyl group because it may be added and removed with facility. Complexes are prepared conveniently from an arene and chromium hexacarbonyl;[41] the activating group is removed by a mild oxidizing agent. The activating effect is similar in magnitude to that of a nitro group.[42]

Arylation of carbanions may be accomplished using chromium tricarbonyl complexes of aryl halides. Tricarbonyl (chlorobenzene) chromium *14* reacts slowly with the anion of isobutyronitrile to form substitution product *15*. Subsequent treatment with iodine in cold ether quantitatively removes the activating group and affords

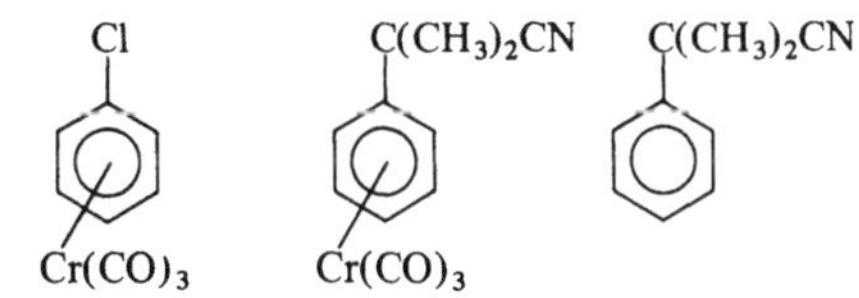

14 *15* *16*

phenylisobutyronitrile (*16*) in 70% yield. Diethyl malonate anion reacts too slowly with *14* in tetrahydrofuran (THF) but changing to the more polar solvent HMPA results in faster conversion; diethyl phenylmalonate is isolated in 51% yield. Failures are encountered; these seem to be associated with primary and secondary carbanions. Chromium complexes resulting from reactions of these anions have acidic centers which can protonate starting carbanion; this leads to side-reactions.[43]

The mechanism of reaction of amines with tricarbonyl (fluorobenzene) chromium in polar aprotic solvents is especially interesting because the chromium tricarbonyl group exerts a steric effect. Nucleophiles attack the ring from the less hindered side *exo* to the chromium center to form σ-complex *17*. Expulsion of fluorine re-

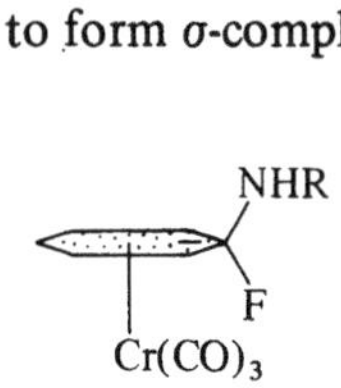

17

quires catalysis by acids to facilitate removal as hydrogen fluoride. But the chromium group hinders the approach of these acids and so general acid catalyzed loss of fluorine is the rate limiting step. In contrast, the rate limiting step in the reactions of *p*-fluoronitrobenzenes in the same solvents where steric factors are not important is addition of a nucleophile to the ring and not the expulsion of fluorine.[42]

3. Ring Cleavage Mechanism: S_N(ANRORC)

Unexpectedly, reactions of some hetaryl halides with amide ion in ammonia do not follow an S_NAr pathway. Although the structures of the products provide no clues that new routes are being followed, the use of isotopically labelled nitrogen atoms reveals unanticipated complexities.

3.1. Pyrimidines

A seemingly simple reaction is the rapid conversion by amide ion in ammonia of 2-halo-4-phenylpyrimidines to the corresponding 2-amino compound in high yield, Eq. (8), R = H. A straightforward interpretation of this result employs an S_NAr mechanism in which amide ion adds to position 2, the center bearing the halogen

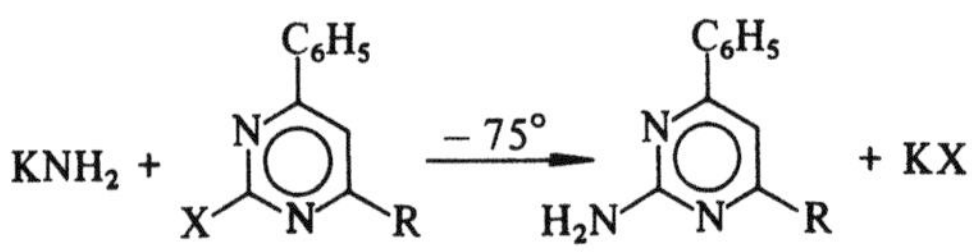

(8)

leaving group, to give a σ-complex. However, this is not the major pathway as revealed by nitrogen-15 labelling studies. Starting material containing label at the two annular nitrogen atoms gives product in which only one of these atoms, N-3, is labelled along with the amino substituent. Therefore, amid ion adds not to position 2 but to position 6. The resulting anionic σ-complex ring-opens with loss of halide ion to form nitrile *18* as an intermediate. This intermediate then cyclizes so as to

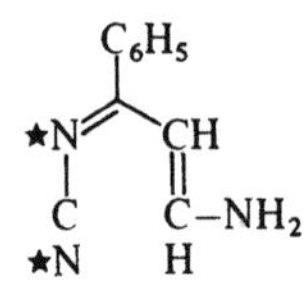

★= ^{15}N

18

incorporate the nitrogen atom from the amide ion into the ring. Shortening the reaction time allows *18* to be isolated. From the amount of scrambling of the label it is estimated that at least 73–88% of the product is formed by ring opening, the remainder probably being formed by the S_NAr route. The extent of the ring cleavage pathway depends on the identity of the halogen atom.[44] Substitution reactions proceeding by way of ring cleavage are designated S_N(ANRORC), the latter symbols being derived from Addition Nucleophile Ring Opening Ring Closure.[45, 46]

Curiously, adding a phenyl group to the starting pyrimidine at position 6 to hinder the addition of amide ion results in a small change in the ratio of pathways when the leaving group is chloride or bromide ion but not when it is fluoride ion. 2-Chloro- and 2-bromo-4,6-diphenylpyrimidines give 70% of substitution product, Eq. (8), R = C_6H_5, by a ring opening route, the remainder is assumed to form by an

S_NAr mechanism. However, the 2-fluoro compound gives no product with scrambled nitrogen label, suggesting that amino product forms only by the S_NAr mechanism.[74)]

Amide ion may also add to position 2 to initiate a ring cleavage mechanism of substitution. Nitrogen labelling studies reveal that all the 4-amino-5-cyano-6-phenyl-

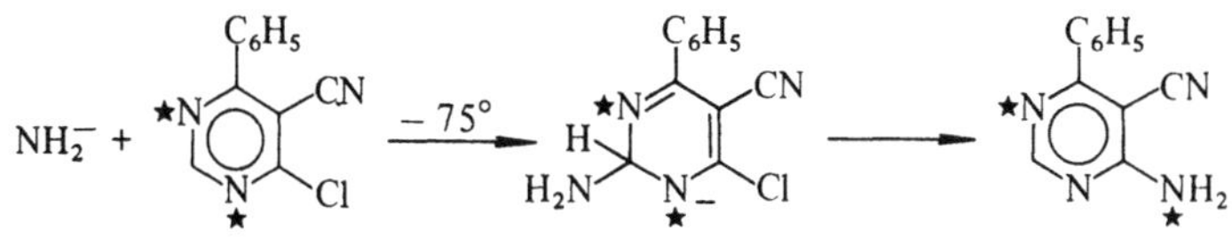

★ = ^{15}N *19* + Cl^- (9)

pyrimidine product formed from the corresponding 4-chloro compound, Eq. (9), arises by an S_N(ANRORC) route. Intermediate σ-complex *19* must be a precursor to acyclic material which cyclizes so as to incorporate the amino group from the solvent into the ring.[48)]

3.2. Pyrazines

The conversion of 2-chloro- to 2-aminopyrazine by amide ion in ammonia does not proceed by an S_NAr mechanism. Starting material labelled at position 1 with nitro-

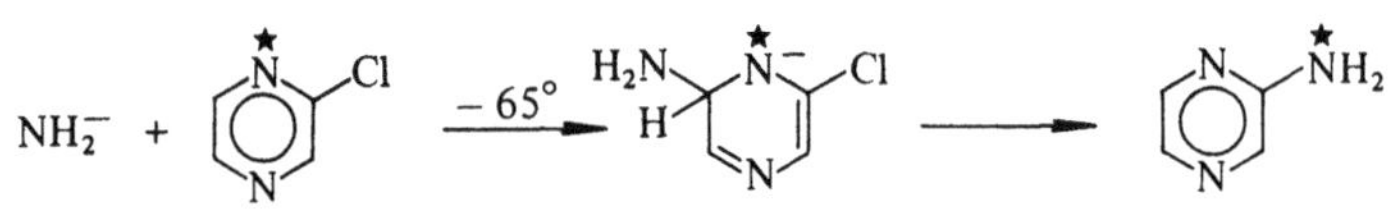

★ = ^{15}N *20* + Cl^- (10)

gen-15 gives rise to product having the label in the amino group instead of the ring, Eq. (10). A ring opening-ring closure sequence is initiated by the addition of amide ion to position 6 to give *20*.[49)]

3.3. Quinazolines

Haloquinazolines also react with amide ion in ammonia to give amino substitution products by an ANRORC process. Nitrogen labelling studies indicate that about half of the 4-aminoquinazoline resulting from 4-chloroquinazoline forms by an S_N(ANRORC) route. Amide ion must add to the unsubstituted 2 position prior to ring opening, Eq. (11).[50)] Similarly, 2-chloro-4-phenylquinazoline gives 2-amino-4-

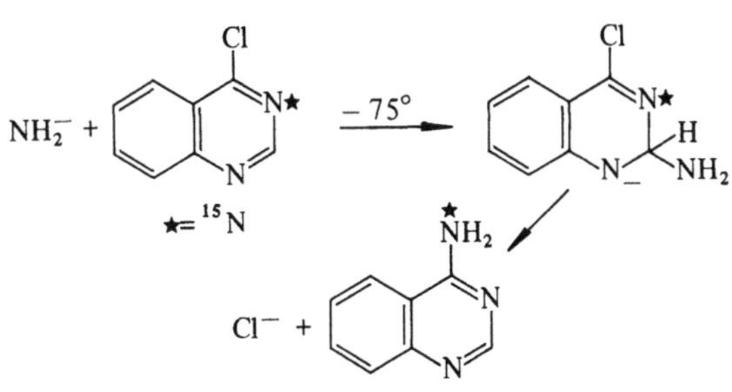

(11)

phenylquinazoline, 70% resulting by an ANRORC process in which amide ion adds to the substituted 4 position to induce ring cleavage, Eq. (12).[47]

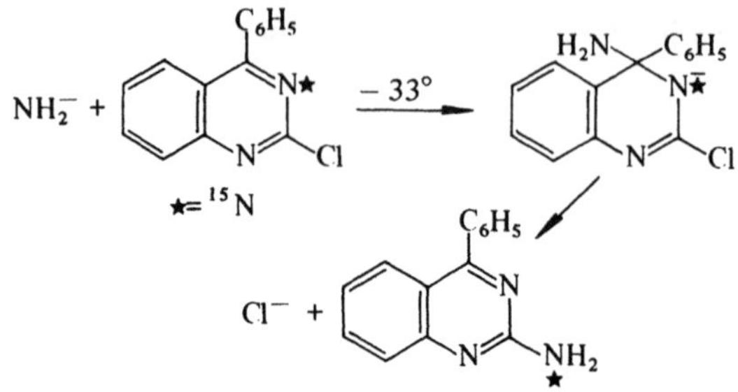

(12)

Using ethanolic ammonia at 160 °C it was shown that, depending on the ammonia concentration, 34–67% of the 2-amino-4-phenylquinazoline produced from the chloro compound arises by a ring opening process. 2-Chloro- and 4-chloroquinazolines also undergo a ring opening substitution pathway in ethanol-ammonia.[47] Thus, the S_N(ANRORC) mechanism is not limited to amide ion in ammonia.

3.4. Isoquinolines

Monoazines may undergo substitution by ring opening routes. Using substrate containing nitrogen-15 it was shown that 55% of the 3-aminoisoquinoline (produced after a 10 minute reaction between 3-bromoisoquinoline and amide ion in ammonia) forms by ring opening. Amide ion must add to position 1 prior to the rearrangement, Eq. (13). In the presence of ethanolic ammonia at 130 °C, 27% of the 3-amino product is formed from bromo compound by ring opening.[51]

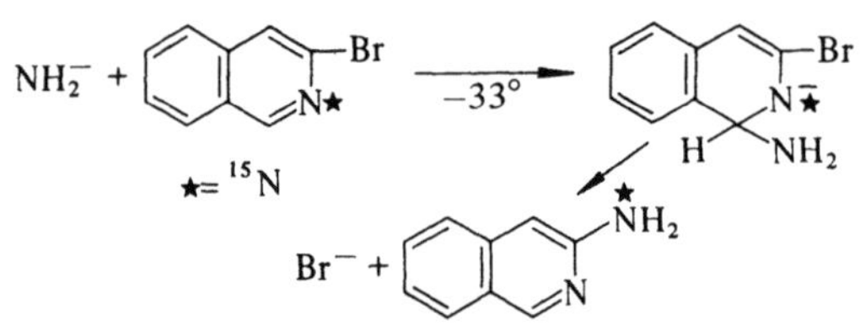

(13)

4. Photosubstitution

Considered are reactions catalyzed by light which do not show a chain mechanism. They differ from photoinitiated radical chain reactions of the $S_{RN}1$ type to be discussed subsequently. Results not included in reviews[52, 53] are examined.

The vagaries of photosubstitution are nicely illustrated by the reactions of 1-methoxy-2-fluoro-4-nitronaphthalene. It undergoes reaction at each of the three substituted positions in the presence of Pyrex filtered light. Substitution takes place at position 1 in the presence of sodium hydroxide in acetonitrile-water, at position

2 using sodium methoxide in methanol and at position 4 when potassium cyanide in acetonitrile-water is employed, Scheme 1.[54)]

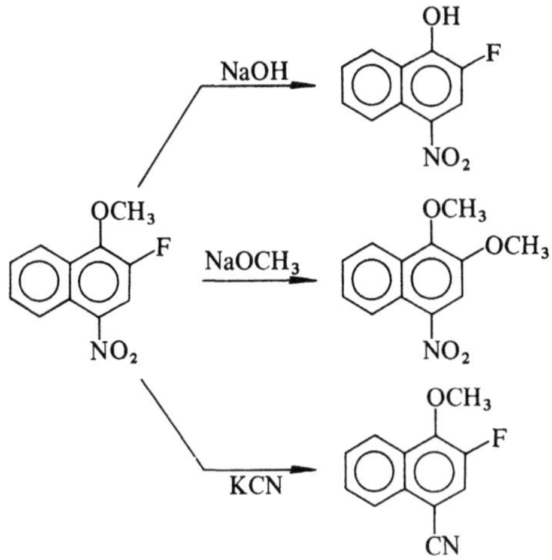

Scheme 1

Orientation rules are just beginning to be formulated: a) Substitution *meta* to a nitro group is favorable. b) Alkoxy and probably other electron donating groups direct to *ortho-para* positions. c) The α-position of polycyclic aromatic compounds is favored. d) Resonance stabilization is important during product formation. The present level of understanding is illustrated with several examples.

The photo conversion of *m*-nitroanisole to *m*-nitrophenolate ion by hydroxide ion, Eq. (14), is sensitized by benzophenone and quenched by oxygen when the solvent is acetonitrile-water. A heavy atom effect is observed, the reaction being prevent-

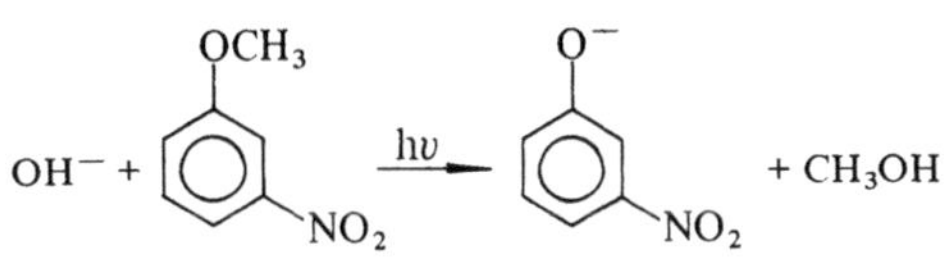

(14)

ed by the presence of sodium bromide. Moreover, flash photolysis reveals that photosubstitution is complete within 20 microseconds and that another component with a lifetime about 40 milliseconds is present as well. This is believed to be the radical anion of starting material.[55)] Similar results are observed for the photo conversion of 3,5-dinitroanisole to 3,5-dinitrophenolate ion by hydroxide ion. "The interaction between excited aromatic compound and nucleophilic reagent leads to the formation of an aromate-nucleophile-complex which is intermediate in the formation of the radical anion and probably also in the formation of the photosubstitution product."[56)] Photosubstitution is likely to start with the lowest $\pi^{\star}$ triplet state of the aromatic substrate.

Evidence has been presented for the formation of a triplet state during the production of 1-methoxyazulene from the nitro compound in the presence of sodium methoxide, Eq. (15). The quantum efficiencies for the reaction vary with wave length and range from a low of < 0.0010 at 546 nm to a high of 0.004 at 254 nm, indicating that intersystem crossing from higher excited singlet states may occur. Photo-

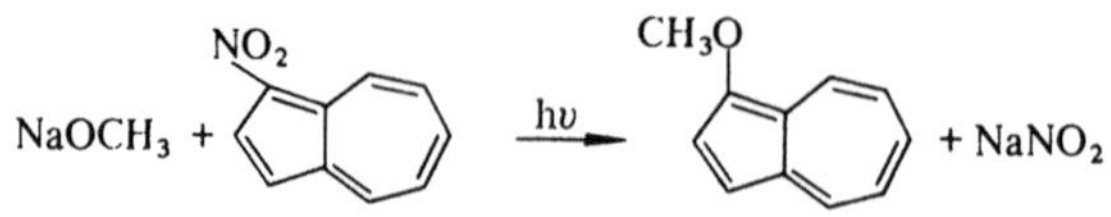

(15)

substitution is retarded in the presence of the triplet quencher 3,3,4,4-tetramethyl-diazetine dioxide and the life-time of the triplet photoexcited state is estimated with the aid of a Stern-Volmer plot to be 10 nanoseconds. The use of nucleophiles such as hydroxide ion, pyridine and cyanate ion which are known to be successful in other photosubstitution reactions failed to give product. However, 1-cyanoazulene was formed in the presence of cyanide ion.[57, 58)]

An unusually large effect of micelles on a quantum yield for a photosubstitution reaction is known. The quantum yield for the conversion of 4-methoxy-l-nitronaphthalene to 4-methoxy-l-cyanonaphthalene by cyanide ion, Eq. (16), is very low, but

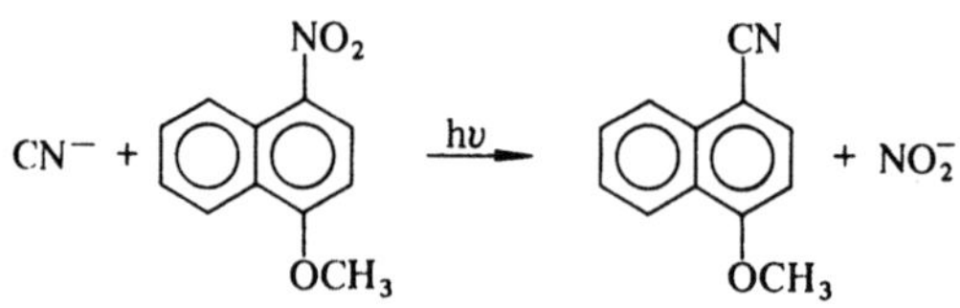

(16)

in the presence of the ionic detergent hexadecyltrimethylammonium chloride this yield is increased by a factor of about 6800. Such a large enhancement is outside the range normally found for micellar effects on ground state reactions and so it is suggested that an excited state effect is in operation. This may be the result of a change in local environment from polar water to less polar hydrocarbon surroundings within the micelles. Unfortunately, much smaller effects of this and other detergents are found for other photosubstitution reactions.[59)]

Photosubstitution reactions have been exploited for their preparative value. Products may be the same as those from thermal reactions but conditions are likely to be milder. Or, entirely new products, often not readily accessible by thermal reactions, may result. The following are typical.

Substitution at a position *meta* to a nitro group occurs readily in the case of 1-fluoro-3-nitronaphthalene irradiated by Pyrex-filtered light. No thermal reaction occurs. Nucleophiles include hydroxide and methoxide ions and methylamine; prod-

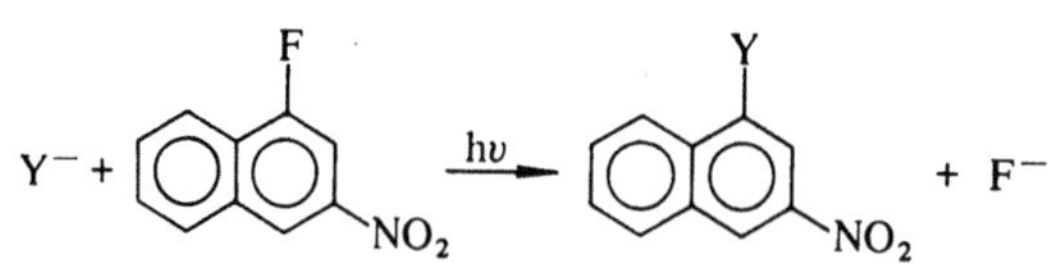

(17)

ucts are isolated in 80–90% yields. Sensitization by benzophenone is efficient and so reactions are believed to involve a triplet state.[54)]

l-Methoxyanthraquinone reacts with ammonia and aliphatic amines in the presence of Pyrex-filtered light to give high yields of 1-amino- and 1-alkylaminoanthraquinones, Eq. (18). Small amounts of an oxidative substitution product 1-amino- or

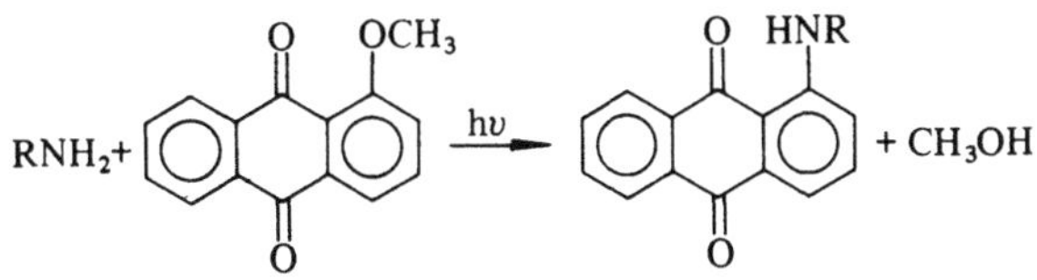

(18)

1-alkylamino-4-methoxyanthraquinone are isolated as well. Increasing the size of the alkyl group results in a decrease in yields, from 96% for the 1-amino to 59% in the case of 1-cyclohexylamino product. Secondary and aromatic amines undergo side reactions preventing their use. Because 1-alkylaminoanthraquinones are valuable as dyes or dye intermediates and thermally induced reactions require severe conditions, the photosubstitution pathway which proceeds at room temperature seems to be useful.[60]

Photolysis of 2-methoxyanthraquinone (*21*) in the presence of ammonia or aliphatic amines serves as a useful way to prepare 1-amino-2-methoxy- and 1-alkyl-

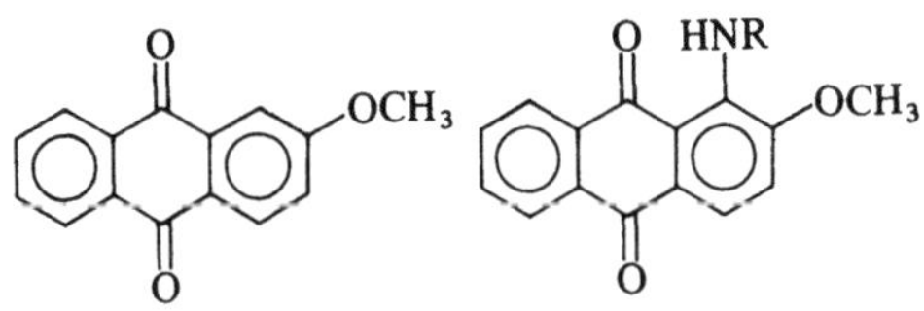

21 *22*

amino-2-methoxyanthraquinones (*22*), compounds which otherwise are accessible only with difficulty. Some 2-amino and 2-alkylamino substitution products are isolated along with the oxidatively substituted materials but their formation can be minimized by adding oxygen.[60]

5. Arenediazonium Ions: S_N1

With this section starts a consideration of substitution reactions in which a bond to the leaving group breaks before a bond to the entering nucleophile forms.

Arenediazonium ion reactions often are complex.[61] Benzenediazonium fluoroborate in 2,2,2-trifluoro ethanol (TFE), for example, decomposes by first-order kinetics to give a mixture consisting largely of the ether (62%) formed by solvolysis and fluorobenzene (34%) resulting from reaction with fluoroborate ion. Addition of pyridine changes the kinetic order and gives rise to additional products.[62] The same diazonium ion solvolyzes faster in TFE than in water. Because water is a better nucleophile than TFE, it is likely that a nucleophile is not involved in the rate limiting step of the reaction.[63]

The mechanisms cited for diazonium ion decomposition continue to remain controversial. For reactions in the absence of strong base, reducing agents and light, a bimolecular mechanism of substitution has been favored since about 1969. The old S_N1 mechanism has been proposed again.[64, 65] In an orbital symmetry allowed reaction linear fragmentation of the diazonium ion leads to nitrogen and singlet

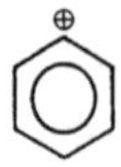

23

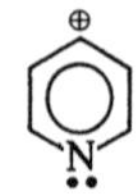

24

phenyl cation *23*.[66] Product determining steps then follow. Important evidence for the S_N1 mechanism is found in experiments dealing with the reactions of benzene-diazonium ion and nitrogen or carbon monoxide under high pressures. Benzene-diazonium ion labelled at nitrogen undergoes exchange with molecular nitrogen to give diazonium ion without label. In the presence of carbon monoxide a benzoic acid derivative results. Since these two gaseous molecules are weak nucleophiles, it is unlikely that they react directly with diazonium ion. More likely is a reaction between nitrogen or carbon dioxide and an intermediate phenyl cation.[65]

Contrary to what might be expected from the electron withdrawing effects of an annular nitrogen atom, extended Hückel theory calculations indicate that 4-pyridyl cation *24* ought to be considerably more stable than a phenyl cation. Interaction between the positive center and the lone electron pair on nitrogen through the intervening σ-bonds is primarily responsible for the added stabilization.[66] Experimental confirmation would be interesting.

Hydroxydediazoniation, Eq. (19), constitutes a useful synthetic reaction of arenediazonium ions. Photolysis rather than the usual thermolysis appears to be a superior way to bring about the transformation. Substituent effects appear to be small but studies have been limited.[53]

$$H_2O + ArN_2^+ \longrightarrow ArOH + N_2 + H^+ \qquad (19)$$

6. Arynes and Hetarynes

Arynes and hetarynes (heterocyclic arynes)[67–72] such as *25* and *26* are named benzyne or dehydrobenzene and 3,4-pyridyne or 3,4-dehydropyridine, respectively.

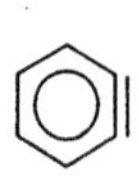

25

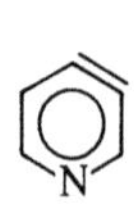

26

They are highly strained and have a singlet ground state.[73, 74] Because such intermediates have two reactive carbon centers they are attractive substrates for cycloaddition reactions. However, little consideration will be given to this important use of arynes and hetarynes because it is outside the scope of this article. Information dealing with the generation of new intermediates is included, although attempts have not yet been made to exploit them in syntheses.

6.1. Dehydrohalogenation

A major route to arynes and hetarynes proceeds by dehydrohalogenating molecules which are deactivated with respect to substitution by an S_NAr mechanism. Frequently employed bases are amide and alkylamide ions, alkyl and aryl lithium reagents and *tert*-butoxide ion. In spite of various claims this approach has not been used successfully to generate a 5-membered hetaryne.[68, 75]

The general mechanism of dehydrohalogenation may be illustrated by considering the reactions of 4-halo-3-methoxymethyl-6-methylpyridazine with amide ion in ammonia, Scheme 2. Amide ion removes a proton at a position adjacent to a halogen;

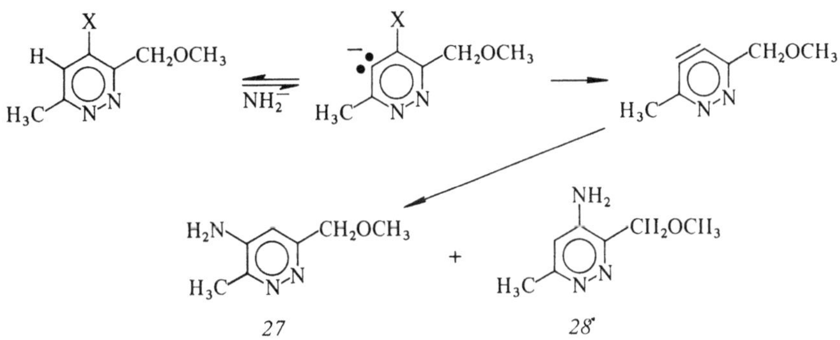

Scheme 2

the resultant anion then eliminates halide ion to generate a pyridazyne. The hetaryne then adds amide ion to either carbon atom of the newly formed unsaturated center to afford new carbanions which are protonated by solvent. Amino substitution products *27* and *28* are isolated. The product ratio is independent of the identity of the halogen leaving group, providing evidence for the formation of a common intermediate, the pyridazyne.[76]

Good nucleophiles are required if an aryne or hetaryne is to be trapped by other than the base employed for its generation. This is an important limitation for syntheses. Among the best nucleophiles are those with sulfur[77, 78] or carbon centers. Isomer formation is another limitation. This is especially severe with *meta* substituted halides which can dehydrohalogenate to give two intermediates which then can react with a nucleophile to form three isomers. However, by carefully selecting a halide so that only one intermediate results and one product isomer is favored, useful syntheses are possible.

3-Methoxybenzyne (*29*), for example, is generated from *o*-chloroanisole and sodium amide in THF. It may be captured by the anion of phenylacetonitrile[79] or cyclohexanone[80] to give arylated material in about 53% yield, Scheme 3. Products, which are useful to the pharmaceutical industry, have groups in a *meta* orientation. By comparison, this approach has limited value for reactions involving *o*-chlorotoluene or *o*-dichlorobenzene and phenylacetonitrile. Mixtures of products resulting from the corresponding 3-substituted benzyne intermediates are difficult to separate.[79]

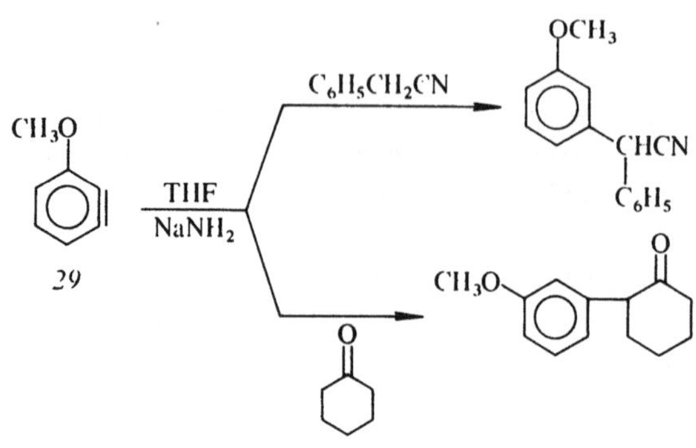

Scheme 3

Intramolecular addition offers one way to direct the position of attack of a nucleophile and prevent isomer formation. Thus, bromopyridine *30* undergoes dehydrobromination in the presence of lithium diethylamide and the resultant pyri-

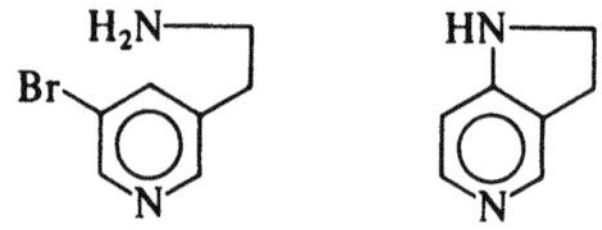

30 *31*

dyne reacts with the attached nucleophilic center to form *31* in 56% yield. The size of the side chain allows addition to only one site and prevents isomer formation.[81)]

A promising development makes use of aprotic solvents, sodium amide and a salt which may be a sodium alkoxide or a sodium enolate. The key to the success of the method lies in the use of the salt which serves to increase the solubility of the amide and perhaps to change its reactivity as well. Sodium amide dissolves slowly in THF but when *tert*-butyl alcohol is added to neutralize some of the amide a homogeneous solution rapidly results. The resultant mixture, termed a "complex base", probably consists of several bases in different states of aggregation. Thus, bromobenzene rapidly reacts with diethylamine in THF containing sodium amide and sodium *tert*-

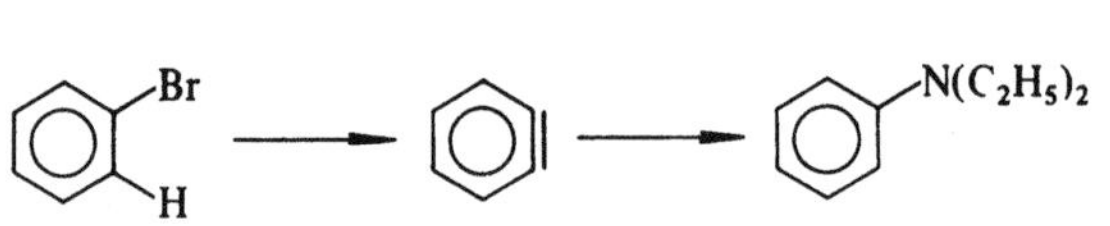

Scheme 4

butoxide. Diethylaminobenzene is formed in a reaction with benzyne in quantative yield, Scheme 4. The reaction fails in the absence of the sodium alkoxide.[82)]

More remarkable are the reactions between ketone enolate ions and arynes generated by a complex base. Products derived using the aprotic solvent system differ considerably from those involving enolate ions in ammonia. In the protic solvent the aryl anion resulting from the addition of an enolate ion to an aryne is protonated and continued reaction is thereby prevented. Such protonation does not occur in the aprotic medium; further reactions occur by intramolecular addition of the aryl anion to a carbonyl group, Scheme 5. Products obtained from acyclic ketones include

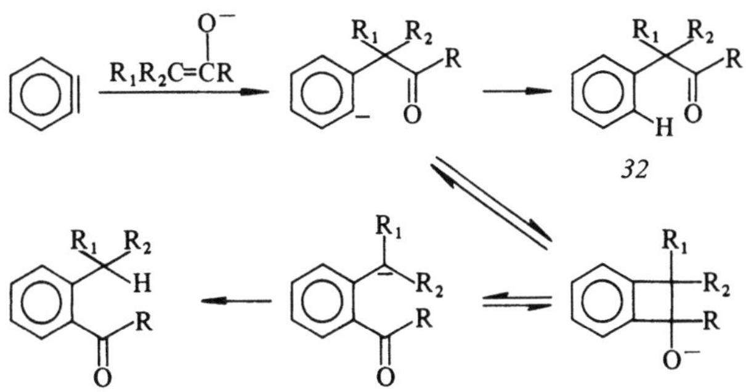

32

33

Scheme 5

arylated ketone *32* and *o*-substituted phenyl ketone *33*. An even broader array of products is obtained with alicyclic ketones. Although difficult separations may be encountered the overall simplicity of the method is an advantage.[82]

When adopting the dehydrohalogenation method in syntheses it is well to remember that halogen migration of the type discussed in connection with S_NAr substitution may be a problem. Halogen scrambling is especially common with bromo compounds.[83, 84]

6.2. From Arenediazonium Ions

One of the simplest routes to benzyne starts with aniline which is converted to benzenediazonium ion. Benzyne is liberated when this ion is deprotonated by acetate base in solvents such as benzene, Eq. (20). Detracting from this approach, however, are competing reactions in which phenyl radicals and phenyl cations are formed. Presently it is possible to minimize the importance of radical chain side reactions and thereby to favor benzyne formation by adding a radical trapping agent such as 1,1-diphenylethylene.[85]

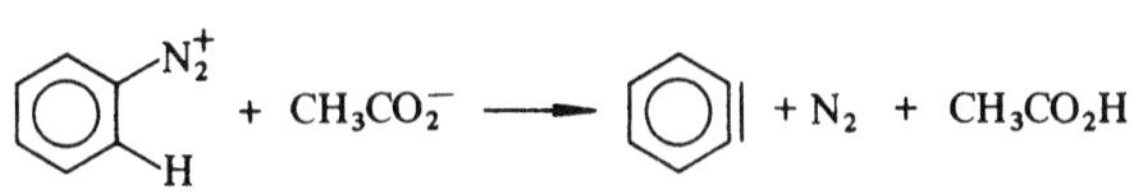

(20)

6.3. Cleavage Reactions

Another major approach to the generation of arynes and hetarynes involves the use of precursors which can be induced by thermal and photochemical means to eliminate small molecules such as CO, CO_2, SO_2 and N_2 along with the desired intermediate. Precursors, for example, include *34*,[86] *35*,[87] and *36*.[88]

A related approach makes use of an N-aminotriazole such as *37*[89] or *38*[90] which on oxidation with lead tetraacetate liberates two equivalents of nitrogen along with a hetaryne. An advantage of this method is that a range of temperatures may be used to generate an intermediate.

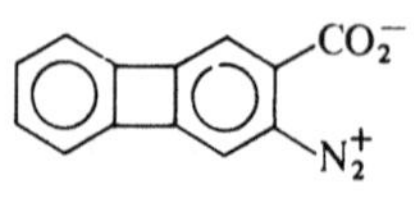
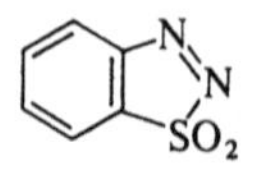
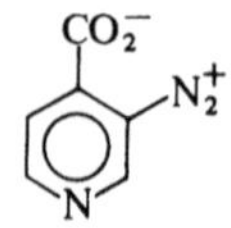

34 *35* *36*

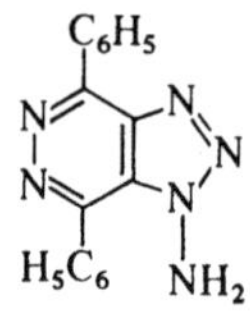
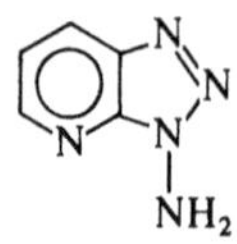

37 *38*

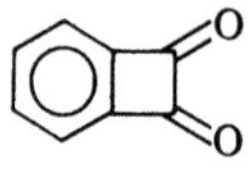
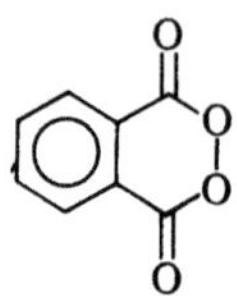

39 *40*

Photochemical routes to arynes are particularly valuable for studies involving low temperatures. For example, benzyne is generated at 8 °K when either benzocyclobutenedione (*39*) or phthaloyl peroxide (*40*) are photolyzed. The benzyne so generated is trapped in a matrix, making it possible to record its infrared spectrum.[91)]

An attempt to generate benzyne in its triplet state has failed. Photolysis in the presence of benzophenone sensitizer converts *40* to its triplet state. But a multistep sequence to benzyne and carbon dioxide allows triplet to singlet intersystem crossing to proceed before the aryne is liberated. It is suggested that triplet benzyne will have to be formed by a concerted reaction of a triplet precursor.[92)]

Methods which do not employ strong bases are especially attractive when the intermediate is to be trapped by reaction with a poor nucleophile or by cycloaddition or "ene" reactions. Illustrative is the reaction between benzenediazonium-2-carboxylate and a furanophane to give a product resulting from two (4+2) cycloaddition reactions with benzyne, Eq. (21).[93)]

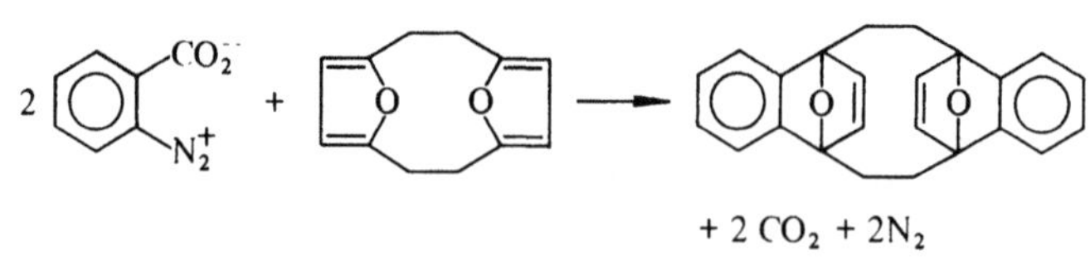

(21)

6.4. Meta and Para Intermediates

There are indications that *meta* and *para* arynes may form. A favorite way to generate such intermediates is by inducing appropriate diazonium carboxylates to eliminate nitrogen and carbon dioxide. Some question exists whether these intermediates have singlet or triplet ground states. The results of MINDO/3 calculations indicate that *m*-benzyne has a singlet ground state which is more stable than heretofore believed. The suggestion that *o*- and *m*-benzynes should have comparable stability is likely to encourage new experimentation with the *meta* intermediate.[94)]

A novel route to *p*-benzyne employs cyclization. Heating *cis*-3-hexene-1,5-diyne in the presence of carbon tetrachloride leads to *p*-dichlorobenzene. *p*-Benzyne is believed to be an intermediate,[95)] Scheme 6.

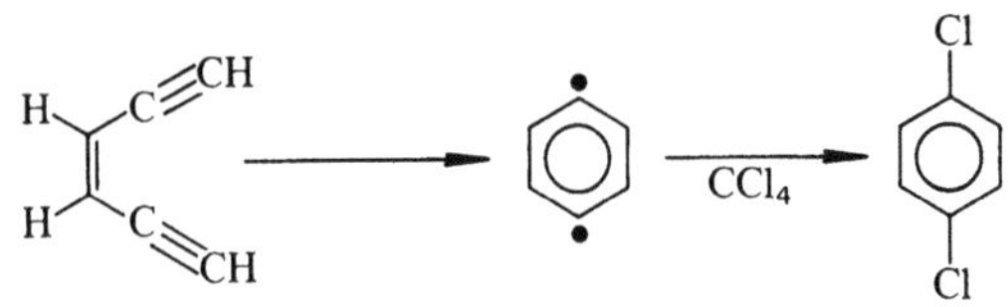

Scheme 6

7. Radical Chain Substitution: $S_{RN}1$

A new and general method of substitution at aromatic carbon has been uncovered. Reactions involve electron transfer steps and the formation of radical anion and radical intermediates. Many of the transformations are unprecedented in aromatic chemistry; they represent major, new sequences which will have considerable value in syntheses. The radical anion method of substitution at aromatic carbon was first reported by Kim and Bunnett in 1970.[96, 97)] The method is related to radical-anion substitution at aliphatic carbon.[98, 99)]

The general mechanism of substitution by the radical anion route can be outlined as follows. A radical chain process is initiated by electron transfer to aromatic substrate ArX to give a radical anion $ArX^{\cdot-}$. The initiating electron may come from various sources such as a metal, a nucleophile, or an electrode. The radical anion then fragments to a carbon radical and a nucleofugic leaving group. Reaction between the aryl or hetaryl radical and a nucleophile gives the radical anion of the substitution product which can transfer an electron to starting material to continue the chain and to give product, Eqs. (22–25). The identity of termination steps is unknown. A wide variety of leavings groups and nucleophiles may be employed. The reaction is designated $S_{RN}1$ which means *S*ubstitution, *R*adical-*N*ucleophile, unimolecular. Unimolecular bond cleavage occurs with the radical anion.[96)]

$$ArX + \text{“}e^-\text{”} \longrightarrow ArX^{\cdot-} \quad (22)$$

$$ArX^{\cdot-} \longrightarrow Ar\cdot + X^- \quad (23)$$

$$Ar\cdot + Y^- \longrightarrow ArY^{\cdot-} \quad (24)$$

$$ArY^{\cdot-} + ArX \longrightarrow ArY + ArX^{\cdot-} \quad (25)$$

7.1. Carbon Nucleophiles

7.1.1. Enolate Ions

A good illustration of the scope of radical chain substitution in terms of the variety of leaving groups which may be employed is found in the reactions of potassium acetonate in ammonia with carbocyclic aromatic compounds. A mixture of substitution products is formed. These include arylacetones, 1-aryl-2-propanols and occasion-

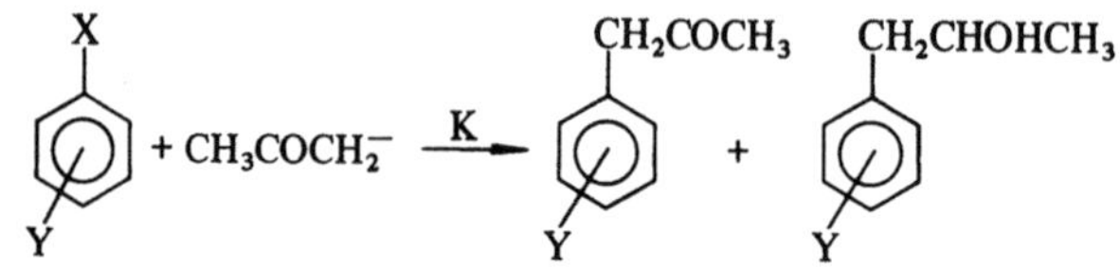

Scheme 7

ally small amounts of 1,1-diarylacetones, Scheme 7. The propanols are formed by reduction of the acetone products. Reaction is initiated by adding potassium metal. Nucleofugic leaving groups include

F, Cl, Br, I, $(CH_3)_3N^+$, C_6H_5O, $(C_2H_5O)_2PO_2$ and C_6H_5S

but not CH_3O. Isolated yields of arylacetone and 1-aryl-2-propanol combined are as high as 71%. The mixture can be oxidized to convert the propanol to an acetone so that only arylacetone is isolated. The use of aryltrimethylammonium ions and aryl diethyl phosphate reactants is especially significant because they are readily prepared from anilines and phenols, respectively. This method of substitution is preferred over an aryne route when the aryne can give rise to a mixture of isomeric products.[100)]

Some leaving groups cannot be employed because cleavage leads not to a radical but to an anion which does not react with a nucleophile to give substitution product. The guiding principle behind the sense of cleavage of the radical anion intermediate is the electronegativity of the leaving group. Cleavage of the bond to the aromatic ring generally occurs so that the more electronegative fragment leaves with a pair of electrons. For example, cleavage to a phenyl anion rather than to a phenyl radical occurs with oxidized sulfur compounds such as diphenyl sulfone and diphenyl sulfoxide. Thioanisole is unsuitable because it undergoes fragmentation at the sulfur-methyl bond. Cyanide ion cannot be induced to leave, owing to faster reduction of the cyano group and/or the aromatic ring.[101)]

Although a radical may be formed on cleavage, it does not necessarily give rise to substitution product. Nucleophiles and electrons both compete for the radical. Reaction with an electron leads to an anion which then is protonated. The net result of the two electron capture sequence is removal of the side chain by reduction. Such a sequence can be used advantageously. Phenols may be dehydroxylated on a preparative scale by converting them to their phosphate esters prior to treatment with an alkali metal in ammonia.[102)]

Many of the $S_{RN}1$ reactions studied to date have employed liquid ammonia as solvent. This is an attractive solvent because it allows weak carbon acids such as acetone to be converted completely to their conjugate bases by reaction with amide ion. Moreover, reaction between ammonia and electrons is slow and alkali metal initiators may be employed. Removal of the solvent, bp –33 °C, following reaction is especially easy.

Photolysis is a particularly effective and convenient way to promote substitution. Arenes containing various leaving groups react with acetone enolate ion at –33° under the stimulation of Pyrex filtered light to give phenylacetone in yields which sometimes are nearly quantitative. The reactivity order in terms of leaving groups is

$$I \sim Br > C_6H_5S \gg Cl > F \gg C_6H_5O.$$

Iodo- and bromobenzene are totally consumed within five minutes. Oxygen provides strong inhibition and demonstrates the chain character of the reactions.[103)]

Little has yet been done to exploit the $S_{RN}1$ mechanism in heterocyclic chemistry. Most of the examples in this area deal with 2-chloroquinoline and enolate ions. 2-Chloroquinoline reacts under photostimulation with lithioacetone to give 2-quinolyl-acetone, Eq. (26). Substitution takes place slowly in the dark unless an electron donor

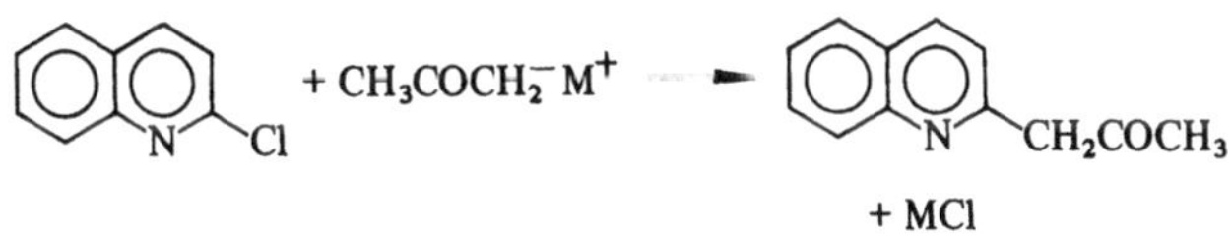

(26)

such as dilithiobenzoylacetone is present as an initiator. Curiously, the presence of excess lithium amide improves the yield.[104)] Substitution is not effectively promoted by lithium metal; a complex mixture of products is formed. However, potassium may be employed successfully with potassium acetonate but 29% of quinoline is produced along with 43% of the substitution product.[105)] The lithium salt of acetophenone also undergoes photostimulated substitution with 2-chloroquinoline and in this instance the presence of lithium amide is not required.[104)]

Ketones which form different enolate ions on deprotonation at the two α positions give rise to mixed substitution products. If anion equilibration is fast with respect to carbon-carbon bond formation, then the product ratio will be governed by two factors: the concentrations of the enolate ions and the selectivity of the capturing radical.

The selectivity of the 2-quinolyl radical is greater than that of the phenyl radical. Consider the results in the case of 3-methyl-2-butanone which gives two enolate ions when deprotonated by amide ion. The less substituted enolate ion is believed to be the major component. When this mixture is allowed to react with the 2-quinolyl

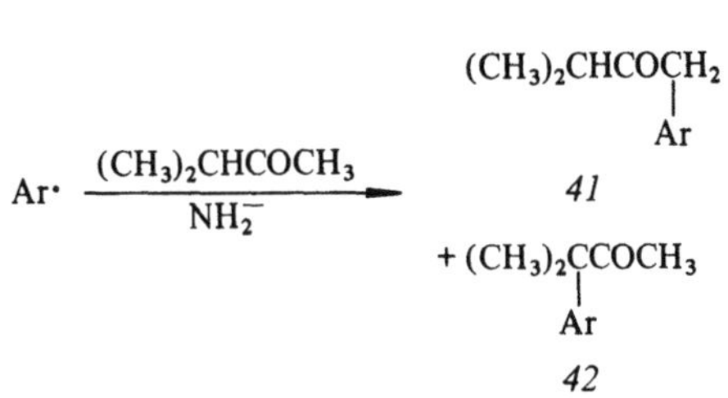

Scheme 8

radical two ketonic products resulting from reactions of both enolate ions are found, Scheme 8. The l-hetaryl ketone *41* is formed in a 4.8 to 1 ratio relative to the 3-hetaryl ketone *42*.[105] In the case of the phenyl radical reacting with the same mixture of enolate ions two ketonic products are produced again, Scheme 8, and the ratio of the corresponding ketones is 9 to 1.[106] In both cases the major product results from reaction with the more abundant enolate ion. But the 2-quinolyl radical shows a stronger preference than the phenyl radical for the less abundant ion. Assuming rapid equilibration of enolate ions it may be concluded that the 2-quinolyl radical is more electrophilic than the phenyl radical and therefore reacts more extensively at the position of greater electron density, *i.e.*, at the more highly substituted carbon. This is understandable if the transition state is polarized so that the radical has anionic

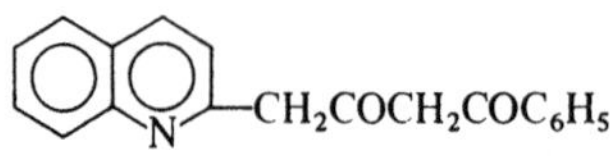

43

character. An annular nitrogen atom is known to stabilize an anion,[107] and to make an aromatic ring a better electron acceptor.[108]

Enolate ions derived from β-dicarbonyl compounds are unreactive toward 2-chloroquinoline even under photolytic conditions.[105] However, the dilithium salt of benzoylacetone does react by an $S_{RN}1$ mechanism with 2-chloroquinoline. The less substituted carbon atom of the dianion is the nucleophilic site; *43* forms.[109]

7.1.2. Nitrile Carbanions

Anions of aliphatic nitriles react by the $S_{RN}1$ route with benzenes containing a variety of leaving groups. Products include the α-phenyl derivative of the nitrile *44*, the alkylbenzene *45* which results from the loss of cyanide ion from *44*, benzene and

$$C_6H_5X + RCH(CN)^-K^+ \xrightarrow{K} \underset{44}{C_6H_5CHR(CN)} + \underset{45}{C_6H_5CH_2R} + C_6H_6 + \text{other products}$$

Scheme 9

other minor components, Scheme 9. Decyanated product results when the radical anion of phenylated nitrile loses cyanide ion to give a radical which is then reduced to a carbanion by reaction with solvated electrons. Protonation of this carbanion by solvent yields the observed alkylbenzene. Decyanation of isolated phenylated nitrile may be carried out in a separate reaction using potassium in ammonia in order to increase the overall yield of alkylbenzene. Although the combined yields of alkylated benzenes never exceeded 50% in the examples explored to date this method has several advantages over the Friedel-Crafts method of alkylation. The position of attachment of the side-chain is limited to the site occupied by the leaving group and is not subject to directing effects of substituents as is the case with electrophilic alkylation. The aliphatic side-chain does not isomerize during the $S_{RN}1$ process but may do so under Friedel-Crafts conditions.[110] The method appears to be especially attractive for heteroaromatic molecules where Friedel-Crafts alkylation fails, owing to deactivation of the heterocyclic compound by reaction between the basic center in the molecule and the acid catalyst.

7.1.3. Other Carbanions

Phenylation of carbanions derived from olefins such as 1,3-pentadiene and 1-(*p*-anisyl) propene and aromatics such as indene and fluorene are possible by the $S_{RN}1$ route in ammonia solvent. From the 1,3-pentadiene a mixture of mono-olefins and dienes is produced along with a small amount of di- and triphenylated material. Hydrogenation of the mixture gives l-phenylpentane in 74% yield.[106] Phenylation of 2- and 4-picolyl anions is conveniently effected in ammonia. When bromo- or iodobenzene is used for this purpose, reaction probably occurs by both $S_{RN}1$ and benzyne routes.[111]

The anion of 2-nitropropane rapidly reacts with 2,4-dinitrochlorobenzene in DMSO or DMF to give a complex mixture of products containing 20% α, 2,4-trini-

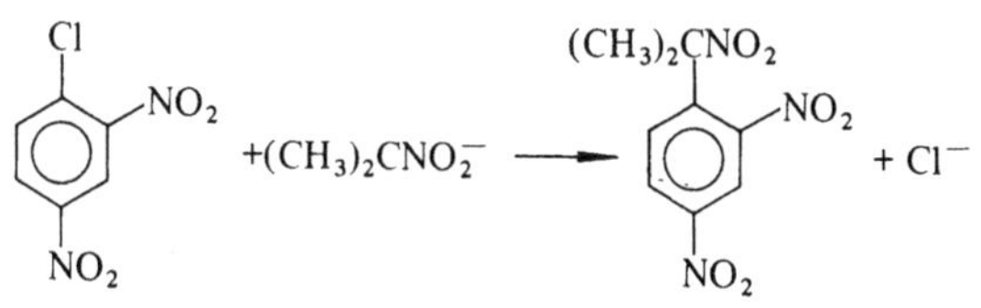

(27)

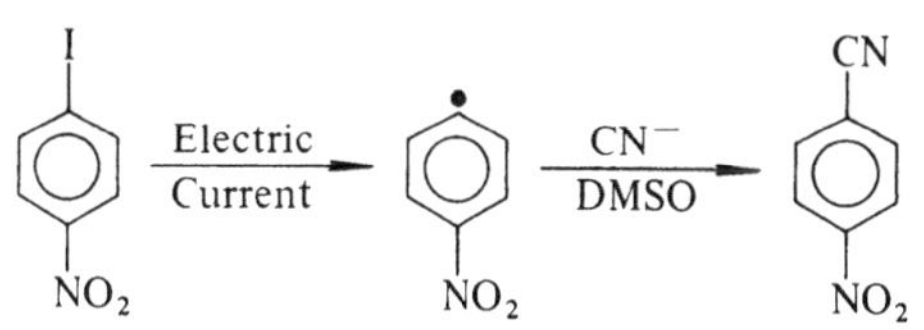

Scheme 10

trocumene, Eq. (27). This product probably arises by trapping of the 2,4-dinitrophenyl radical by the 2-nitro-2-propyl anion.[112]

The conversion of *p*-iodonitrobenzene to *p*-cyanonitrobenzene is an example of an electrochemically initiated $S_{RN}1$ reaction, Scheme 10. Cyanide ion traps the intermediate *p*-nitrophenyl radical.[113]

7.2. Sulfur Nucleophiles

An excellent preparation of diaryl sulfides from aryl iodides and arenethiolate ions has been devised. Photolysis brings about rapid formation of the sulfide by a radical-chain process. Yields of isolated product seldom fall below 75%, Eq. (28). This ap-

$$ArI + Ar'S^- \xrightarrow{h\nu} ArSAr' + I^- \qquad (28)$$

pears to be the method of choice for the preparation of aryl sulfides containing groups which retard nucleophilic substitution by the ionic S_NAr method. It also appears to be superior to preparations which employ copper catalysts.[114] The same method may also be utilized to prepare *bis* sulfides from disubstituted benzenes. An unusual feature of this preparation is that the monosubstitution product is not an intermediate. Instead, the radical anion of the monosubstitution product *46*, produced by addition of thiophenoxide ion to a substituted phenyl radical, Scheme 11,

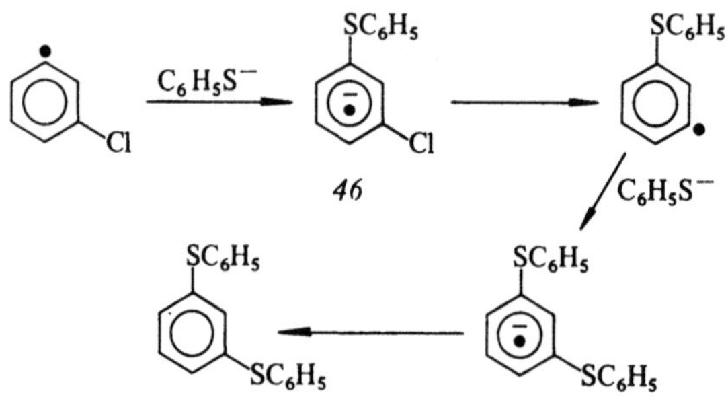

Scheme 11

undergoes elimination to afford a second phenyl radical which in turn gives rise to the final product.[115] The monosubstitution product would be an intermediate if the radical anion lost an electron rather than chloride ion.

4-Bromoisoquinoline may be made to react with benzenethiolate ion to give 4-phenylthioisoquinoline in methanol at 147 °C by a radical chain route. The interesting feature is that the radical ion process requires methoxide ion for initiation. In the absence of methoxide ion substitution takes place at a slower rate by an S_NAr route. Formerly, the presence of methoxide ion in substitution reactions involving thiolate ions in methanol was avoided studiously in order to prevent a competing substitution by the methoxide ion. Although methoxide ion is required for the radical chain process to occur, only traces of substitution product containing the methoxy group are present, demonstrating that benzenethiolate ion is superior to methoxide ion in trapping the intermediate 4-isoquinolyl radical,[116] Scheme 12.

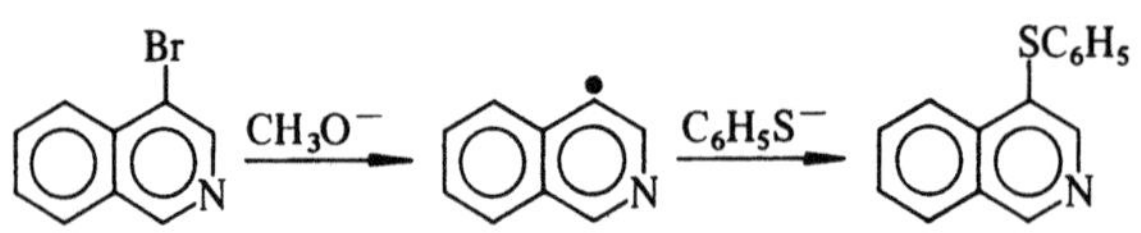

Scheme 12

7.3. Nitrogen Nucleophiles

Few generally useful methods have been available for transforming phenols to anilines in high yields. The $S_{RN}1$ method now offers such a possibility. A phenol is converted to its aryl diethylphosphate ester prior to treatment with potassium in

$$ArOH \xrightarrow[(C_2H_5O)_2POCl]{} ArOPO(OC_2H_5)_2 \xrightarrow[KNH_2]{K} ArNH_2$$

Scheme 13

ammonia, Scheme 13. The intermediate aryl radical derived from the ester is trapped by amide ion. It is suggested that phenols having alkyl, aryl, alkoxy, aryloxy and enolizable ketone substituents will react only at the phenolic center but that nitrated or halogenated compounds are expected to undergo additional reactions.[117] It is well to remember that amide ion can compete with other nucleophiles for intermediate radicals in ammonia. High concentrations of this lyate ion should generally be avoided when other nucleophiles are present.

Ammonia rather than amide ion serves as a nucleophile during the photolytic aminodehalogenations of pyrimidines. 4-Halo-6-substituted pyrimidines give the corresponding amino compound on photolysis in quartz vessels, Scheme 14. The

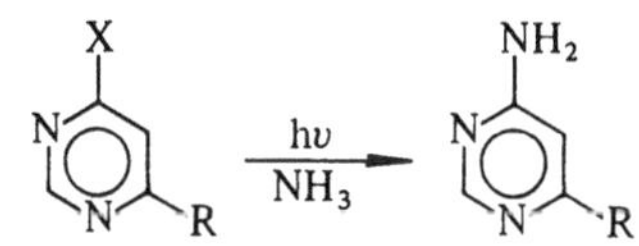

Scheme 14

reactivity order is I > Br > Cl. A radical chain mechanism is postulated because no reaction occurs in the absence of light and a nitroxide inhibits photolytic substitution.[118]

7.4. Phosphorus Nucleophiles

Arylphosphonate esters are conveniently prepared from aryl iodides and potassium dialkyl phosphites on photolysis in ammonia, Eq. (29). Isolated yields of products containing methyl, methoxy, trifluoromethyl or halogen substituents range from 87 to

$$ArI + (RO)_2PO^-K^+ \xrightarrow[h\upsilon]{} ArPO(OR)_2 + KI \qquad (29)$$

96%.[119] These are likely to be chain reactions as opposed to the non-chain reaction of a phenyl radical with trimethyl phosphite which also produces a phenylphosphonate ester.[120]

8. Metal Ion Catalysis

Metal ions can accelerate nucleophilic substitution reactions as well as bring about transformations which would not occur in their absence. Often the mechanism of catalysis is obscure. Copper salts are among the oldest and most useful.[121, 122]

The best available method for converting nonactivated aryl halides into aryl alkyl ethers employs copper(I). Aryl bromides and iodides react with copper(I) alkoxides in pyridine to give the ethers in high yield, Eq. (30). Although substitution

$$ArX + ROCu \longrightarrow ArOR' + CuX \quad (30)$$

takes place at room temperature, reactions are more conveniently carried out at 115 °C. The mechanism is not established but $S_{RN}1$ and oxidative addition of copper(I) to the aromatic ring are possibilities.[123)]

A newly emerging class of reactions is oxidative substitution. An aromatic molecule reacts with a nucleophile to give a substitution product, a proton and two electrons or their equivalent. The aromatic molecule is first oxidized by a metal ion (or even an anode[124)]) to a radical cation intermediate which then is trapped and converted to product. Consider the reaction of alkylbenzenes with copper(II) chloride and peroxydisulfate ion. The peroxy compound converts toluene, for example, into its radical cation which is trapped in a reaction with the copper salt. Aromatization then gives products which are mostly *o*- and *p*-chlorotoluene, Scheme 15. In the

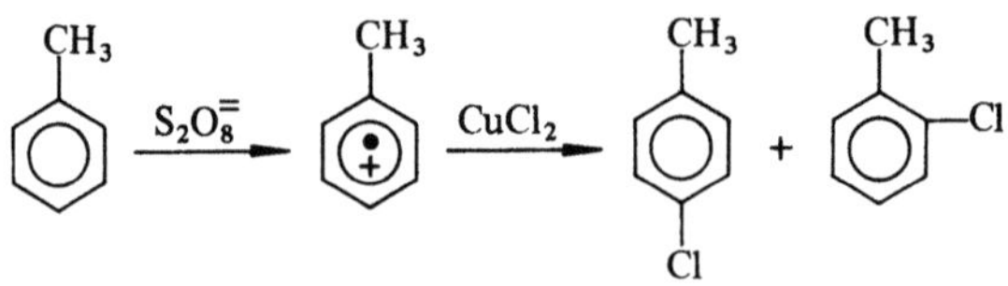

Scheme 15

absence of the copper salt a benzyl radical forms instead. Copper(II) salts therefore appear to be useful reagents to trap radical cations.[125, 126)] Other oxidative halogenation reactions with copper salts are known.

2-Alkoxynaphthalenes are converted to l-chloro- or l-bromo-2-alkoxynaphthalenes in near quantitative yield using copper(II) chloride or bromide. l-Alkoxynaphthalene affords about 70% l-alkoxy-4-bromonaphthalene with copper(II) bromide. Similarly, anisole gives a mixture of *o*- and *p*-haloanisoles. The absence of *meta* isomer is believed to provide evidence for the intermediacy of a radical cation of anisole.[127)]

Introduction of a trifluoroacetoxy group onto a benzene ring is achieved using cobalt(III) trifluoroacetate in trifluoroacetic acid. Electron transfer from benzene to cobalt(III) produces a cation radical and cobalt(II). The benzene radical cation

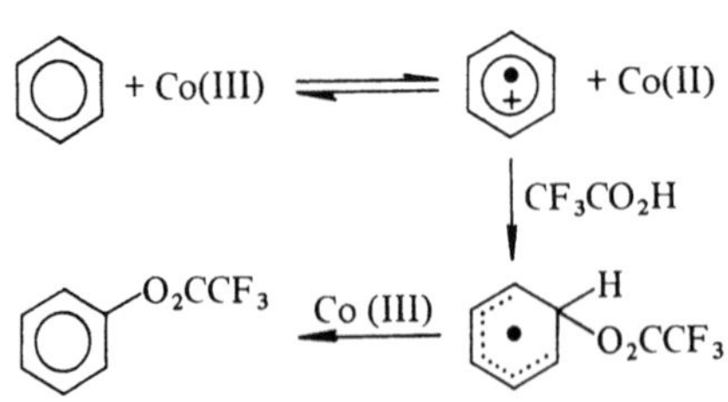

Scheme 16

then reacts with the solvent and the resultant radical aromatizes on undergoing oxidation by cobalt(III), Scheme 16. With benzene the exothermic reaction is complete in about 30 minutes at room temperature. Chlorobenzene gives a mixture of *o*- and *p*-chlorophenyl trifluoroacetates in 18 and 35% yields, respectively. The *ortho* to *para* product ratio in this case is similar to that observed for electrophilic but not for homolytic substitution. However, no substitution products involving ring positions are found with toluene. Instead, side-chain substituted and polymeric materials result.[128)]

Generally it is difficult to distinguish between electrophilic addition to an aromatic ring and one electron removal. The first process gives a cationic and the second a radical cationic intermediate. Substituent effects are expected to influence both kinds of reactions in the same way.[129)] The same can be said for nucleophile versus one electron addition to an aromatic substrate to yield an anion and a radical anion, respectively. Reactions proceeding by a pair of one electron transfer reactions to give radical ion intermediates rather than by processes which yield purely ionic intermediates may be more common than is realized.[129–132)] New research will provide an answer.

Acknowledgement: Prof. J. F. Wolfe kindly provided unpublished results. Prof. L. S. Helmick made helpful comments about the manuscript.

9. References

1) Miller, J.: Aromatic nucleophilic substitution. New York: Elsevier 1968.
2) Bernasconi, C. F.: in MTP international review of science. Aromatic compounds. Series 1, Vol. 3, p. 33. (H. Zollinger, Ed.). Baltimore: University Park Press 1973.
3) Mathieu, J., Allais, A., and Vallas, J.: Angew. Chem., *72*, 71 (1960).
4) Ritchie, C. D.: Accts. Chem. Res., *5*, 348 (1972).
5) Krueger, J. H., Sudbury, B. A., and Blanchet, P. F.: J. Amer. Chem. Soc., *96*, 5733 (1974).
6) Kice, J. L., Rogers, T. E., and Warheit, A. C.: J. Amer. Chem. Soc., *96*, 8020 (1974).
7) Brown, H. C., and Krishnamurthy, S.: J. Amer. Chem. Soc., *95*, 1669 (1973).
8) Arnett, E. M., Small, L. E., McIver Jr., R. T., and Miller, J. S.: J. Amer. Chem. Soc., *96*, 5638 (1974).
9) Brauman, J. I., Olmstead, W. N., and Lieder, C. A.: J. Amer. Chem. Soc., *96*, 4030 (1974).
10) Abramovitch, R. A., and Newman Jr., A. J.: J. Org. Chem., *39*, 3692 (1974).
11) Bunnett, J. F.: J. Chem. Soc., 4717 (1954).
12) Sanger, F.: Biochem. J., *45*, 563 (1949).
13) Zoltewicz, J. A., and Sale, A. A.: J. Org. Chem., *35*, 3462 (1970).
14) Zoltewicz, J. A., and Sale, A. A.: J. Org. Chem., *36*, 1455 (1971).
15) Orvik, J. A. and Bunnett, J. F.: J. Amer. Chem. Soc., *92*, 2417 (1970).
16) Crampton, M. R., Ghariani, M. A. El, and Khan, H. A.: Tetrahedron, *28*, 3299 (1972).
17) Crampton, M. R., and Khan, H. A.: J. Chem. Soc., Perkin II, 710 (1973).
18) Zoltewicz, J. A., Helmick, L. S., Oestreich, T. M., King, R. W., and Kandetzki, P. E.: J. Org. Chem., *38*, 1947 (1973).
19) Zoltewicz, J. A., and Helmick, L. S.: J. Amer. Chem. Soc., *94*, 682 (1972).

20) Geerts, J. P., Plas, H. C. van der, and Veldhuizen, A. van: Rec. Trav. Chim. (Pays-Bas), *92*, 1232 (1973).
21) Zoltewicz, J. A., and Helmick, L. S.: J. Org. Chem., *38*, 658 (1973).
22) Pedersen, C. J., and Frensdorff, H. K.: Angew. Chem. Internat. Ed. Engl., *11*, 16 (1972).
23) Christensen, J. J., Eatough, D. J., and Izatt, R. M.: Chem. Revs., *74*, 351 (1974).
24) Cook, F. L. Bowers, C. W., and Liotta, C. L.: J. Org. Chem., *39*, 3416 (1974).
25) Sam, D. J., and Simmons, H. E.: J. Amer. Chem. Soc., *96*, 2252 (1974).
26) Cima, F. Del, Biggi, G., and Pietra, F.: J. Chem. Soc., Perkin II, 55 (1973).
27) Parker, A. J.: Chem. Revs., *69*, 1 (1969).
28) Abramovitch, R. A., and Newman, A. J., Jr.: J. Org. Chem., *39*, 2690 (1974).
29) Knudsen, R. D., and Snyder, H. R.: J. Org. Chem., *39*, 3343 (1974).
30) Bradshaw, J. S., and Hales, R. H.: J. Org. Chem., *36*, 318 (1971).
31) Hales, R. H., Bradshaw, J. S., and Pratt, D. R.: J. Org. Chem., *36*, 314 (1971).
32) Bradshaw, J. S., Chen, E. Y., Hales, R. H., and South, J. A.: J. Org. Chem., *37*, 2051 (1972).
33) Bradshaw, J. S., South, J. A., and Hales, R. H.: J. Org. Chem., *37*, 2381 (1972).
34) Zoest, W. J. van, and Hertog, H. J. den: Rec. Trav. Chim. (Pays-Bas), *93*, 166 (1974).
35) Barker, J. M., Coutts, I. G. C., and Huddleston, P. R.: Chem. Commun., 615 (1972).
36) Bunnett, J. F.: Accts. Chem. Res., *5*, 139 (1972).
37) Dehmlow, E. V.: Angew. Chem. Internat. Ed. *13*, 170 (1974).
38) Dockx, J.: Synthesis, *1973*, 441.
39) Makosza, M., Jagusztyn-Grochowska, M., Ludwikow, M., and Jawdosiuk, M.: Tetrahedron *30*, 3723 (1974).
40) Makosza, M., and Ludwikow, M.: Angew. Chem. Internat. Ed., *13*, 665 (1974).
41) Marr, G., and Rockett, B. W.: J. Organomet. Chem. *79*, 347 (1974).
42) Bunnett, J. F., and Herman, H.: J. Org. Chem., *36*, 4081 (1971).
43) Semmelhack, M. F., and Hall, H. T.: J. Amer. Chem. Soc., *96*, 7091 (1974).
44) Kroon, A. P., and Plas, H. C. van der: Rec. Trav. Chim. (Pays-Bas), *93*, 111 (1974).
45) Valk, J. de, and Plas, H. C. van der: Rec. Trav. Chim. (Pays-Bas), *90*, 1239 (1971).
46) Plas, H. C. van der, and Koudijs, A.: Rec. Trav. Chim. (Pays-Bas), *92*, 711 (1973).
47) Kroon, A. P., and Plas, H. C. van der: Rec. Trav. Chim. (Pays-Bas), *93*, 227 (1974).
48) Valk, J. de, and Plas, H. C. van der: Rec. Trav. Chim. (Pays-Bas), *92*, 471 (1973).
49) Lont, P. J., Plas, H. C. van der, and Verbeck, A. J.: Rec. Trav. Chim. (Pays-Bas), *91*, 949 (1972).
50) Valk, J. de, Plas, H. C. van der, Jansen, F., and Koudijs, A.: Rec. Trav. Chim. (Pays-Bas), *92*, 460 (1973).
51) Sanders, G. M., Dijk, M. van, and Hertog, H. J. den: Rec. Trav. Chim. (Pays-Bas), *93*, 198 (1974).
52) Stenberg, V. I., and Dutton, D. R.: Tetrahedron, *28*, 4635 (1972).
53) Matsuura, T., and Omura, K.: Synthesis, *1974*, 173.
54) Lammers, J. G., and Lugtenberg, J.: Tetrahedron Letters, *1973*, 1777.
55) Heijer, J. den, Spee, T., Gunst, G. P. de, and Cornelisse, J.: Tetrahedron Letters, *1973*, 1261.
56) Gunst, G. P. de, and Havinga, E.: Tetrahedron, *29*, 2167 (1973).
57) Lok, C. M., Lugenburg, J., Cornelisse, J., and Havinga, E.: Tetrahedron Letters, *1970*, 4701
58) Lok, C. M., Boer, M. E. den, Cornelisse, J., and Havinga, E.: Tetrahedron, *29*, 867 (1973).
59) Hautala, R. R., and Letsinger, R. L.: J. Org. Chem., *36*, 3762 (1971).
60) Griffiths, J., and Hawkins, C.: J. Chem. Soc., Perkin I, 2283 (1974).
61) Zollinger, H.: Accts. Chem. Res., *6*, 335 (1973).
62) Buuri, P., Loewenschuss, H., Zollinger, H., and Zwolinski, G. K.: Helv. Chim. Acta., *57*, 395 (1974).
63) Burri, P., Wahl, Jr., H. G., and Zollinger, H.: Helv. Chim. Acta, *57*, 2099 (1974).
64) Swain, C. G., Sheats, J. E., Gorenstein, D. G., Harbison, K. G., and Rogers, R. J.: Tetrahedron Letters, *1974*, 2973.
65) Bergstrom, R. G., Wahl Jr., G. H., and Zollinger, H.: Tetrahedron Letters, *1974*, 2975.
66) Gleiter, R., Hoffmann, R., and Stohrer, W. D.: Chem. Ber., *105*, 8 (1972).

67) Hoffmann, R. W.: in Chemistry of acetylenes. P. 1063, (Viehe, H. G., Ed.). New York: M. Dekker 1969.
68) Hoffmann, R. W.: Dehydrobenzene and cycloalkynes. New York: Academic Press 1967.
69) Fields, E. K., and Meyerson, S.: Advan. Phys. Org. Chem., *6*, 1 (1968).
70) Zoltewicz, J. A.: in International review of science. Aromatic compounds. Series 2, Vol. 3, (H. Zollinger, Ed.). London: Butterworths, in press.
71) Kauffmann, T., and Wirthwein, R.: Angew. Chem. Internat. Ed. *10*, 20 (1971).
72) Hertog, H. J. den, and Plas, H. C. van der: in Chemistry of acetylenes. P. 1149, (Viehe, H. G., Ed.). New York: M. Dekker 1969.
73) Haselbach, E.: Helv. Chim. Acta., *54*, 1981 (1971).
74) Adam, W., Grimison, A., and Hoffmann, R.: J. Amer. Chem. Soc., *91*, 2590 (1969).
75) Bie, D. A. de, Plas, H. C. van der, and Geurtsen, G.: Rec. Trav. Chim. (Pays-Bas), *90*, 594 (1971).
76) Klinge, D. E., Plas, H. C. van der, and Koudijs, A.: Rec. Trav. Chim. (Pays-Bas), *93*, 201 (1974).
77) Zoltewicz, J. A., and Nisi, C.: J. Org. Chem., *34*, 765 (1969).
78) Huisgen, R.: in Organometallic chemistry. P. 38, (H. Zeiss, Ed.). ACS Monographs, No. 147. New York: Reinhold, 1960.
79) Kametani, T., Kigasawa, K., Hiiragi, M., Aoyama, T., and Kusama, O.: J. Org. Chem., *36*, 327 (1971).
80) Kametani, T., Noguchi, S., Agata, I., Aono, T., Kigasawa, K., Hiiragi, M., Hayasaka, T., and Kusama, O.: J. Chem. Soc. (C), 1047 (1971).
81) Kauffmann, T., and Fischer, H.: Chem. Ber., *106*, 220 (1973).
82) Caubere, P.: Accts. Chem. Res., *7*, 301 (1974).
83) Plas, H. C. van der, Bie, D. A. de, Geurtsen, G., Reinecke, M. G., and Adickes, H. W.: Rec. Trav. Chim. (Pays-Bas), *93*, 33 (1974).
84) Hertog, H. J. den, and Buurman, D. J.: Rec. Trav. Chim. (Pays-Bas), *92* 304 (1973).
85) Cadogan, J. I. G., Murray, C. D., and Sharp, J. T.: Chem. Commun., *1974*, 901.
86) Losey, E. N., and LeGoff, E.: J. Org. Chem., *38*, 3812 (1973).
87) Wittig, G., and Hoffmann, R. W.: Chem. Ber., *95*, 2718 (1962).
88) Kramer, J., and Berry, R. S.: J. Amer. Chem. Soc., *94*, 8336 (1972).
89) Gilchrist, T. L., Gymer, G. E., and Rees, C. W.: Chem. Commun. *1973*, 819.
90) Fleet, G. W. J., Fleming, I., and Philippides, D.: J. Chem. Soc., *C*, 3948 (1971).
91) Chapman, O. L., Mattes, K., McIntosh, C. L., Pacansky, J., Calder, G. V., and Orr, G.: J. Amer. Chem. Soc., *95*, 6134 (1973).
92) Luibrand, R. T., and Hoffmann, R. W.: J. Org. Chem. *39*, 3887 (1974).
93) Kapicak, L. A., and Battiste, M. A.: Chem. Commun. *1973*, 930.
94) Dewar, M. J. S., and Li, W-K.: J. Amer. Chem. Soc., *96*, 5569 (1974).
95) Bergman, R. G.: Accts. Chem. Res., *6*, 25 (1973).
96) Kim, J. K., and Bunnett, J. F.: J. Amer. Chem. Soc., *92*, 7463 (1970).
97) Kim, J. K., and Bunnett, J. F.: J. Amer. Chem. Soc., *92*, 7464 (1970).
98) Russell, G. A.: Essays on free radical chemistry. Chem. Soc. Publ. No. 24, p. 271 (1970).
99) Kornblum, N., Boyd, S. D., and Ono, N.: J. Amer. Chem. Soc., *96*, 2580 (1974) and references cited therein.
100) Rossi, R. A., and Bunnett, J. F.: J. Amer. Chem. Soc., *94*, 683 (1972).
101) Rossi, R. A., and Bunnett, J. F.: J. Amer. Chem. Soc., *96*, 112 (1974).
102) Rossi, R. A., and Bunnett, J. F.: J. Org. Chem., *38*, 2314 (1973).
103) Rossi, R. A., and Bunnett, J. F.: J. Org. Chem., *38*, 1407 (1973).
104) Hay, J. V., Hudlicky, T., and Wolfe, J. F.: J. Amer. Chem. Soc., *97*, 374 (1975).
105) Hay, J. V., and Wolfe, J. F.: J. Amer. Chem. Soc., *97*, 3702 (1975).
106) Rossi, R. A., and Bunnett, J. F.: J. Org. Chem., *38*, 3020 (1973)
107) Zoltewicz, J. A., and Sale, A. A.: J. Amer. Chem. Soc., *95*, 3928 (1973).
108) Wiberg, K. B., and Lewis, T. P.: J. Amer. Chem. Soc. *92*, 7154 (1970).
109) Wolfe, J. F., Greene, J. C., and Hudlicky, T.: J. Org. Chem., *37*, 3199 (1972).
110) Bunnett, J. F., and Gloor, B. F.: J. Org. Chem., *38*, 4156 (1973).

111) Bunnett, J. F., and Gloor, B. F.: J. Org. Chem., *39*, 382 (1974).
112) Russell, G. A., Norris, R. K., and Panek, E. J.: J. Amer. Chem. Soc., *93*, 5839 (1971).
113) Bartok, D. E., Danen, W. C., and Hawley, M. D.: J. Org. Chem., *35*, 1206 (1970).
114) Bunnett, J. F., and Creary, X.: J. Org. Chem., *39*, 3173 (1974).
115) Bunnett, J. F., and Creary, X.: J. Org. Chem., *39*, 3611 (1974).
116) Zoltewicz, J. A., and Oestreich, T. M.: J. Amer. Chem. Soc., *95*, 6863 (1973).
117) Rossi, R. A., and Bunnett, J. F.: J. Org. Chem., *37*, 3570 (1972).
118) Bie, D. A. de, Plas, H. C. van der, and Geurtsen, B.: J. Chem. Soc., Perkin I, 1363 (1974).
119) Bunnett, J. F., and Creary, X.: J. Org. Chem., *39*, 3612 (1974).
120) Fu, J-J. L., and Bentrude, W.: J. Amer. Chem. Soc., *94*, 7710 (1972).
121) Bacon, R. G. R., and Hill, H. A. O.: Quart. Revs., *19*, 95 (1965).
122) Normant, J. F. N.: Synthesis, *1972*, 63.
123) Whitesides, G. M., Sadowski, J. S., and Lilburn, J.: J. Amer. Chem. Soc., *96*, 2829 (1974).
124) Eberson, L., and Nyberg, K.: Accts. Chem. Res. *6*, 106 (1973)
125) Ledwith, A., and Russell, P. J.: Chem. Commun., *1974*, 291.
126) Ledwith, A., and Russell, P. J.: Chem. Commun., *1974*, 959.
127) Bansal, S. R., Nonhebel, D. C., and Mancilla, J. M.: Tetrahedron, *29*, 993 (1973).
128) Kochi, J. K., Tang, R. T., and Bernath, T.: J. Amer. Chem. Soc., *95*, 7114 (1973).
129) Kochi, J. K.: Accts. Chem. Res., *7*, 351 (1974).
130) Kochi, J. K.: Tetrahedron Letters, *1974*, 4305.
131) Fendler, E. J., Fendler, J. H., Arthur, N. L., and Griffin, C. E.: J. Org. Chem. *37*, 812 (1972).
132) Shein, S. M., and Khmelinskaya, A. D.: Zh. Organ. Kh., *8*, 1252 (1972).

Received February 10, 1975

Formation and Reactions of Aminyloxides

Prof. Dr. Hans Günter Aurich and Dr. Wolfram Weiss

Fachbereich Chemie der Philipps-Universität Marburg

Dedicated to Prof. Dr. K. Dimroth on the occasion of his 65th birthday

Contents

I. Introduction

Porphyrexide (*1*) the first organic aminyloxide (nitroxide)[a] was isolated by Piloty[1] in 1901, one year after the discovery of triphenylmethyl. 13 years later Wieland prepared diphenylaminyloxide (*2a*: R = R' = C_6H_5)[2]. The development of ESR spectroscopy made this class of radicals a subject of growing interest. The research was more concentrated to cyclic and acyclic aliphatic aminyloxides such as *3* and *2b*: R = R' = $C(CH_3)_3$ respectively. The pioneering work was done by the research groups of Rozantsev in UdSSR and Rassat in France in the beginning of the sixties.

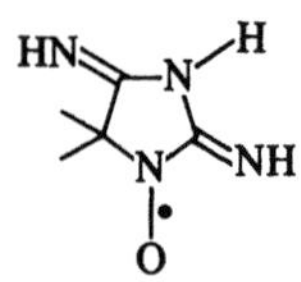

1

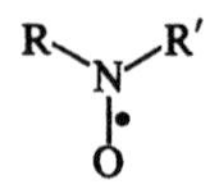

2

R bzw. R' = H, CH_3
prim.–, sec.–, tert–alkyl,
aryl, NR_2,OR,SiR_3, PR_3

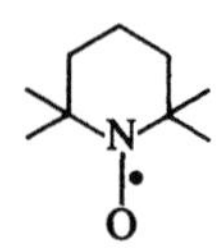

3

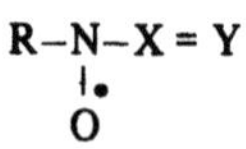

4

X	C–R'	C–R'	C–R'	C–R'	C–X
Y	N–R''	$N^{\oplus}$–R'' ↓ $O_{\ominus}$	N–OH	O	O

X	N	N	N	C–R'	N
Y	N–R''	O	C–R''_2	C–R''_2	$N^{\oplus}$–R'' ↓ $O_{\ominus}$

At present many aminyloxides of different structure are known, for instance aminyloxides *2* with primary, secondary or tertiary alkyl groups, hydrogen, amino- or alkoxy groups and substituents containing elements of higher periods. Many aminyloxides *4* having conjugated multiple bonds are also known.

Rozantsev described the simple methods of preparation and the reactions of stable aminyloxides in his book "Free Nitroxyl Radicals"[3] and in a review article[4] that was published some years ago.

The present work intends to give a summary of the more complex reactions leading to aminyloxides which have been studied in recent years, and of the different possibilities of reactions of aminyloxides resulting from the considerably enlarged number of types of these radicals at present.

---

[a] According to IUPAC rules (C–81.2) these radicals should be named aminyloxides or aminoxyls although the name nitroxide is used extensively.

II. The Delocalization of the Unpaired Electron in the Aminyloxide Group

Compared to aminyl radicals the aminyloxides are radicals of relatively low energy. This is caused by the delocalization of the unpaired electron between the nitrogen and oxygen atom, as described by the two mesomeric formulas *A* and *B*. Molecular

A *B*

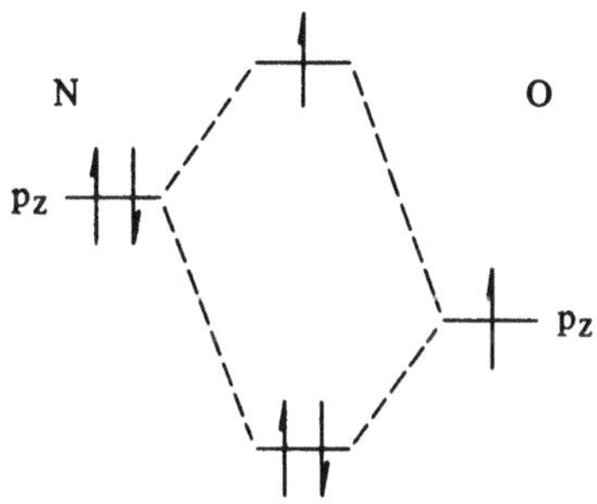

orbitals formed from the atomic orbitals of nitrogen and oxygen have two of the p_z-electrons in the bonding π-orbital the third one in the antibonding π-orbital, yielding a net bonding by one electron. Accordingly, nitrogen and oxygen are connected by a 1 1/2-bond. This was confirmed by the position of the IR- frequencies[5)].

For *3* and 4-hydroxy-2,2,6,6-tetramethylpiperidyloxide the energy of delocalization was estimated to be 30.4 kcal/Mol[6)]. The gain in energy is responsible for the easy formation of aminyloxides in many cases and for the extraordinary stability of such radicals which are prevented from secondary reactions by high activation energy. Thus di-tert-alkylaminyloxides as for instance *3* or *2b*: $R = R' = C(CH_3)_3$ may be isolated and stored for months or years. On the other hand aminyloxides containing primary alkyl groups are so unstable that they usually may be detected only by continuous generation in the ESR-cell (Flow system, photolysis, electrolysis). The coupling constants determined from the ESR spectra indicate the coupling of the unpaired electron with the nuclei which have a non zero spin.

The following coupling constants for nitrogen a^N are characteristic for the different types of aminyloxides.

Dialkylaminyloxides	14–17 G
Alkylarylaminyloxides	11–14 G
Diarylaminyloxides	9–11 G
Acylaminyloxides	6,5–8 G
Amidinyl-N-oxides	7,5–10 G

Amidinyl-N.N'-dioxides	7,0–7,5 G
Alkoxy-alkylaminyl-oxides	24–28 G
Alkoxy-arylaminyloxides	13–15 G

Deviations may be caused by steric effects or substituent effects.

The coupling constant a^N is not an exact measure of the spin density at nitrogen atom, as there is a small contribution of spin density at the oxygen atom to a^N. On the other hand the coupling constant of ^{17}O a^O is also influenced by the spin density at nitrogen atom. Although a^O was measured for some dialkylaminyloxides (a^O = 18.5 – 19.5 G)[7–12] a unique statement about distribution of spin density between nitrogen and oxygen is very difficult even in such simple systems as dialkylaminyl-oxides. In earlier publications values between 0.2 to 0.8[13] are discussed for the spin density at nitrogen atom ρ_N, more recent works conclude that $\rho_N \approx \rho_O$ ($0.55 \leqslant \rho_N \leqslant 0.79$ and $0.53 \leqslant \rho_O \leqslant 0.30$[9b]; $0.4 \leqslant \rho_N \leqslant 0.54$ and $0.6 < \rho_O \leqslant 0.46$[7]).

A further difficulty is that the geometry of the aminyloxide group of the radicals in solution is not known exactly. Generally it is assumed that the aminyloxide group is pyramidal in dialkylaminyloxides whereas it is planar in arylaminyloxides.

Studying ^{17}O-labeled radicals Cohen and Hoffman[9a] came to the conclusion that in solution the deviation from planarity is similar for di-tert-butyl-aminyloxide (*2b*) and 2,2,6,6-tetramethylpiperidyloxide (*3*). On the other hand in the solid state *2b* was found to be planar[14] whereas for *3* an "out of plane" angle of 21° was determined for the

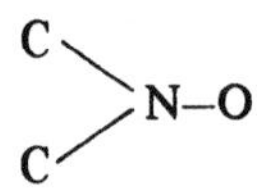
group[15].

Nevertheless the coupling constants a^N may be used as an approximate measure for the spin density at nitrogen in a more qualitative manner. In aminyloxides containing functional groups conjugated to the radical center a^N is lowered clearly by delocalization of the unpaired electron into these groups. The a^O values of ^{17}O-labeled diaryl-aminyloxides (17.2 – 17.5 G[12]) which are only slightly lowered compared to those of dialkylaminyloxides, indicate that the spin density at oxygen is not reduced very much. Thus the delocalization into aryl groups causes mainly reduction of spin density at the nitrogen atom.

The values of a^N are clearly solvent-dependent. Polar solvents enhance the spin density at nitrogen atom favoring structure *A*. In many cases there is a good correlation between a^N and solvent-parameters such as E_T-values[16–18]. Accordingly ^{17}O-labeled 2,2,6,6-tetramethylpiperidon-4-yloxid shows a decrease of a^O with increasing polarity of solvent[7].

Substituent effects by functional groups work in the same manner[19]. The extremely low value a^N = 9.3 G of the bistrifluoromethylaminyloxide[20] compared to 2*b* (a^N = 15.4 G) is caused, at least partly, by the strong inductive effect of the electronegative trifluoromethyl groups shifting spin density towards the oxygen atom.

III. Routes for Formation of Aminyloxides

The following scheme shows the principal types of reaction paths leading to aminyloxides. The more complex ways of formation of aminyloxides may also be derived from these principal reactions.

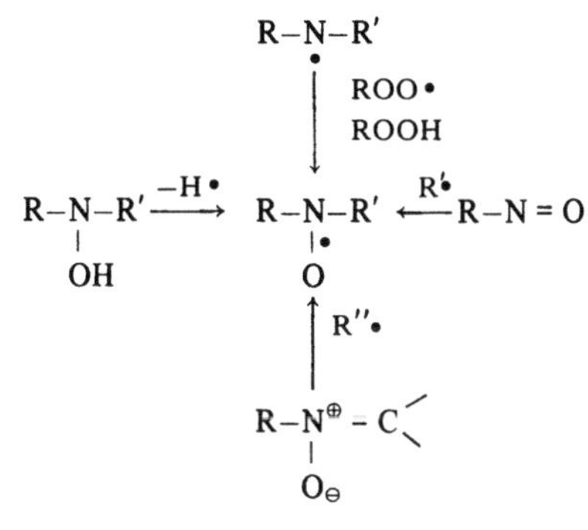

1. Aminyloxides from Hydroxylamines

The most simple method for preparation of aminyloxides is the dehydrogenation of the corresponding hydroxylamines with any of several oxidizing agents[21]. The easy formation of aminyloxides from hydroxylamines is caused by the relatively low energy of the aminyloxide function. The OH bond strength for N-hydroxy-2,2,6,6-tetramethylpiperidine and N-hydroxy-2,2,6,6-tetramethylpiperidon-4 was found to be 69.6 and 71.8 kcal/Mol respectively, whereas for comparable oximes values of 80–85 kcal/Mol were determined[22]. Thus in reaction of hydroxylamines *5* and nitrile oxides *6*, N.N'-dihydroxyamidines *7* could be isolated only in few cases, but even with strong exclusion of oxygen their oxidation products, the radicals *8*, were detected in every case[23]. These radicals accordingly exist as N^2-hydroxyamidinyl-N^1-oxides and not as tautomeric iminyloxides. The oxidizing agent is the nitrile oxide *6* itself being reduced to the corresponding aldoxime. In most cases the radicals *8* give further reactions (see below pp. 96/97). On the other hand reaction of amines and nitrile oxides leads to the corresponding amide oximes in very good yields, no radical being detectable in the reaction mixture[23]. Clearly the different behavior of *7* and the amide oximes towards the weakly oxidizing agent *6* reflects the easy formation of the aminyloxide group from hydroxylamine.

Tautomers of hydroxylamines may also be oxidized easily. Amidin-N^2-oxides as the nitrone form of N^1-hydroxyamidines have been oxidized with different agents[24].

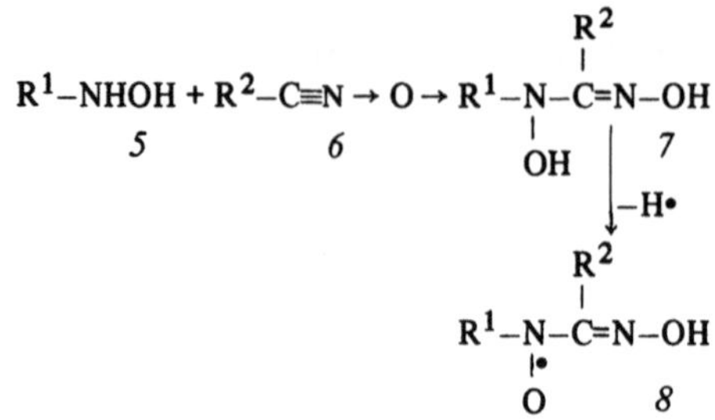

The addition product of tert-butylhydroxylamine (*5a*) and acetylene dicarbonic dimethylester *9*, the nitrone *10*, which could be detected only at low temperature by NMR, was converted to the vinylaminyloxide *11* by direct oxidation of the reaction mixture[25].

Finally 2,3-diaryl-5-iminoisoxazolines as cyclic tautomers of β-cyanovinylhydroxylamines yielded the acyclic β-cyanovinylaminyloxides after oxidation with lead dioxide[26].

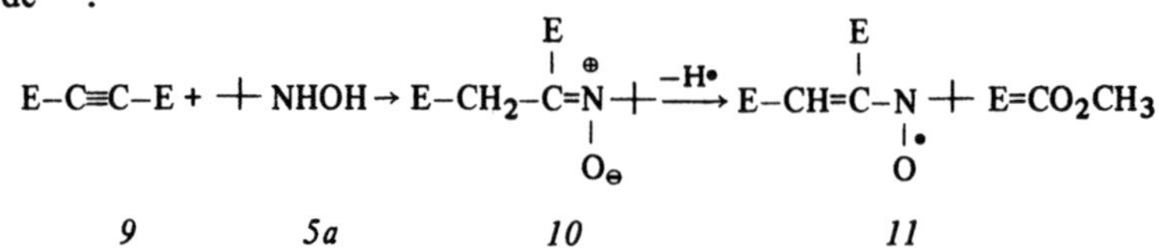

The oxidation of deprotonated hydroxylamines occurs even more easily of course. In alkaline medium in presence of Cu^{2+}-ions the anion of dibenzylhydroxylamine is oxidized even by air[27]. In absence of oxygen under basic conditions dibenzylhydroxylamine is converted to the corresponding aminyloxide by m-dinitrobenzene[28], with nitrobenzene no reaction occurs[29].

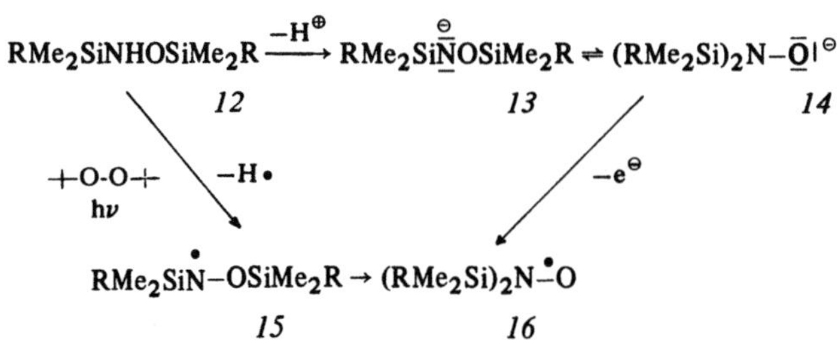

The anions *13* of the N.O-substituted hydroxylamines *12* are in equilibrium with the rearranged anions *14*. Oxidation by oxygen or electrolytically yields the corresponding aminyloxides *16*. Abstraction of hydrogen from *12* also yields aminyloxides *16* which must be formed by rearrangement of the intermediate radical *15*[30].

2. Aminyloxides from Amines, Hydrazines and Tetrazenes

A very favorable method of generation of disubstituted aminyloxides is the direct oxidation of secondary amines, much more easily available than the corresponding hydroxylamines. Using this method many aminyloxides have been prepared by oxidation with hydrogen peroxide in presence of tungstate, molybdate or vanadate, alkaline hydrogen peroxide, perbenzoic acids, hydroperoxides, lead dioxide or other oxidizing agents[31)].

For instance ^{17}O-labeled aliphatic aminyloxides were obtained from amines either by oxidation with ^{17}O-labeled hydrogen peroxide[7)] or were formed in solution by irradiation in presence of ^{17}O-labeled oxygen gas and di-tert-butylperoxide[12)] or azo-bis-isobutyro-nitrile[9a)]. Photolysis of N-chloramines in benzene containing oxygen yields aminyloxides[32)].

Formation of diarylaminyloxides from tetraarylhydrazines by hydroperoxides or lead dioxide in dioxan has been found long ago[33)]. Thermolysis of hydrazines in toluene in the presence of ^{17}O-labeled oxygen made ^{17}O-labeled diarylaminyloxides detectable[12)]. Aryl-tert-butylaminyloxides have also been obtained in the presence of oxygen from aryl-tert-butylaminyl radicals generated by thermolysis or photolysis of corresponding hydrazines or tetrazenes[34)]. Thermolysis of O-tert-butyl-N-aryl-peroxycarbamates leads to N-tert-butoxy-N-aryl-aminyl radicals, but in presence of tert-butylhydroperoxide N-tert-butoxy-N-aryl-aminyloxides are formed[35)].

It seems obvious that all these different ways of formation of aminyloxides lead through intermediate aminyl radicals. But this is not generally true. Reaction of secondary amines with hydrogen peroxide/tungstate is assumed to proceed through hydroxylamine which is oxidized in a subsequent reaction step to aminyloxide[36)]. In the reaction of diphenylamine and peroxy radicals no isotope effect was observed. Therefore, it was suggested that a primary complex *17* was formed which then reacts with a second peroxy radical yielding diphenylaminyloxide (*2a*), either directly or through diphenylaminyl intermediate (*18*)[37)].

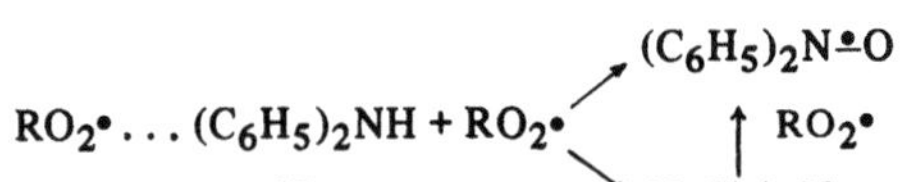
2a *17* *18*

On the other hand it seems certain that oxidation of hydrazines and tetrazenes occurs through intermediate aminyl radicals. Unambiguous evidence for direct reaction of aminyl radical and oxygen could be obtained only for reaction of 2,2,6,6-tetramethylpiperidyl. This radical is at once converted to corresponding ^{17}O-labeled aminyloxide *3* when ^{17}O-oxygen is allowed to diffuse into the cooled solution whereas under the same conditions in absence of oxygen it is relatively stable[8)].

In all other cases the possibility may not be excluded, that the reaction proceeds through peroxy radical intermediates, which may be formed by hydrogen abstraction from solvent and subsequent reaction of the resulting alkyl radicals with oxygen.

3. Aminyloxides from Nitroso Compounds

Nitroso compounds are very important starting materials for generation of aminyloxides. A great number of different types of aminyloxides has been obtained by spin trapping[38]. A short-lived radical generated in presence of a nitroso compound (spin trap) forms a relative stable aminyloxide (spin adduct). Principally it is also

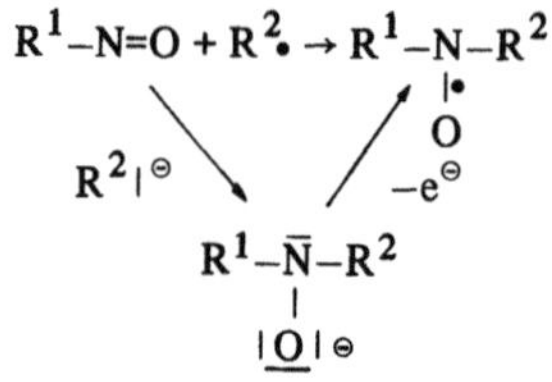

19

possible that a nucleophile adds to nitroso compound in a non-radical manner yielding adduct *19* which is then oxidized to aminyloxide (by air or excess of nitroso compound respectively). Thus detection of an aminyloxide does not definitely allow the conclusion that an intermediate radical $R^2\cdot$ exists[39]. The most important nitroso compounds being used as spin traps are: 2-methyl-2-nitrosopropane, $[D_9]$-2-methyl-2-nitrosopropane[40], 2-methyl-2-nitrosobutanone-3[41]. 1-nitrosoadamantane[42], nitrosocyclopropane[43], nitrosobenzene and some substituted aromatic nitroso compounds[44].

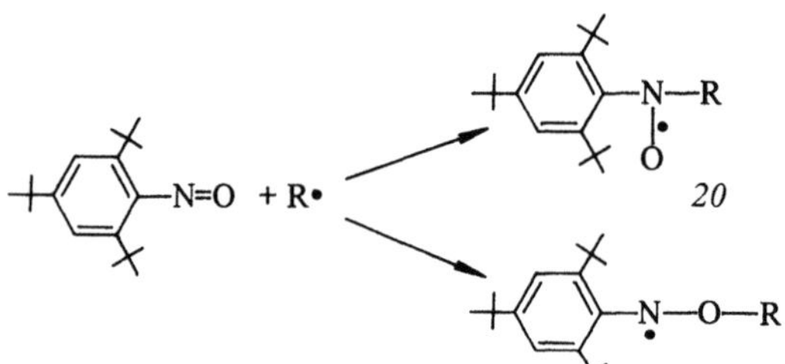

20

21

The sterically hindered 2,4,6-tri-tert-butyl-nitrosobenzene shows a very interesting behavior giving aminyloxides *20* and O-substituted aminyls *21* respectively, dependent on size of the radical being trapped[45]. Thus primary alkyl radicals give spin adducts *20*, tertiary radicals give spin adducts *21*, whereas secondary alkyl radicals yield a mixture of *20* and *21*. The sterically more crowded triethylsilyl- and triethylstannyl radicals also form adduct *21*. Alkyl radicals generated from alkylbromides and tributylstannylhydride add to the stable 2,2-di-tert-butyl-1-nitrosoethylene *22*

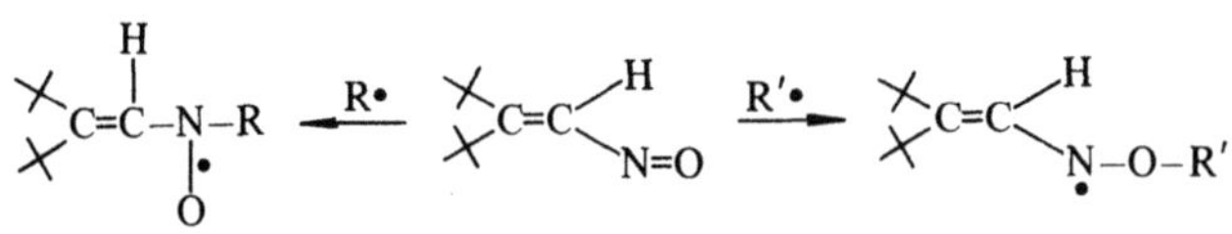

23 *22* *24*

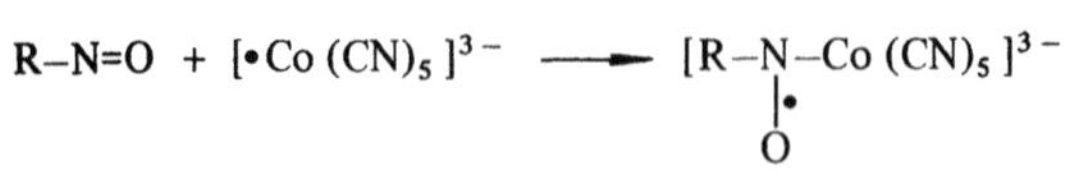

25 *26*

giving aminyloxide *23*[46a]. On the other hand tert-butyl radicals formed by irradiation of 2,2-azoisobutane surprisingly yield spin adducts *24*[46].

Different types of carbon radicals as well as nitrogen,[47] oxygen[47f, g, 48], sulfur[47f, g] and phosphor[49] radicals have been trapped by nitroso compounds.

A very interesting spin-trap reaction is the reaction of paramagnetic potassium-pentacyano-cobaltate (II) *25* and aliphatic or aromatic nitroso compounds leading to pentacyano-cobalt-(III)aminyloxide radical anions *26*[50].

Formation of aminyloxides by irradiation of nitroso compounds is a special case of spin trapping. Fragmentation of photochemically excited nitroso compound gives nitric oxide and radical R·, the latter being trapped by unchanged nitroso compound[51]. Homolysis of nitroso compound may also occur thermally[52]. Thus for instance the dissociation energy of the C–N-bond of 2-methyl-2-nitroso-propane was determined as 34 kcal/Mol[53]. For comparison C–N-bond dissociation energy in tert-butyl-amine is 75 kcal/Mol.

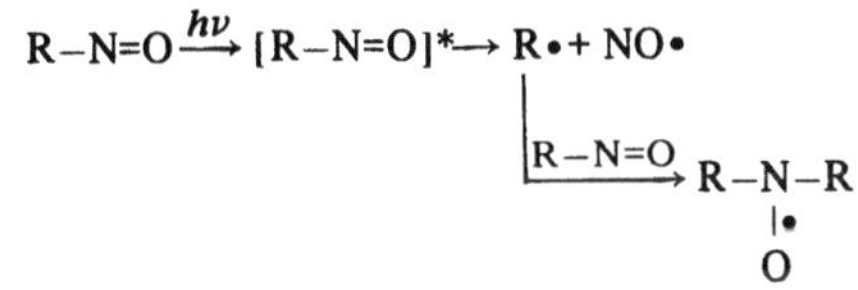

The appearance of free radicals R· in this process has been proved by de Boer[54]. Photolyzing 1-cyclopropyl-1-nitroso-ethane *27* he obtained aminyloxide *30*. In this way it is shown that the initially formed 1-cyclopropylethyl radical *28* by ring opening converts to Δ^3-pentenyl radical *29* which then adds to unchanged *27*.

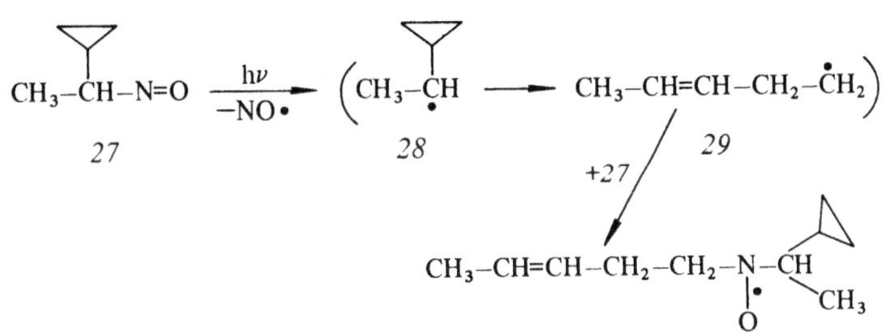

27 *28* *29* *30*

An example of intramolecular spin trapping was studied by A. Rassat[55]. Heating the bicyclic compound *31* in cyclohexane leads to dinitroso compound *32*. Elimination of nitric oxide occurs by irradiation of the refluxing solution, and the resulting radical suffers intramolecular spin trapping giving aminyloxide *33*.

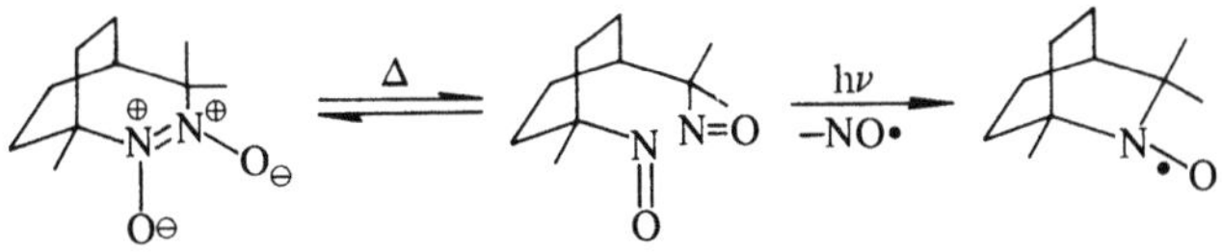

31 *32* *33*

There are, of course, many different methods for the generation of the primary radicals which are then trapped by the nitroso compound. Most frequently hydrogen abstraction from solvent or a specific radical precursor by the readily available tert-butoxy radical[48d)] has been applied. But there are also some other interesting radical producing reactions giving abstraction of hydrogen from solvent thus yielding aminyloxides in different ways[48a, 56)].

Addition of radicals to alkenes giving secondary radicals that are trapped terminally by nitroso compound should be mentioned further. The latest known example is addition of tert-butoxy radicals to styrene, the resulting β-tert-butoxyalkyl radical being trapped by 2-methyl-2-nitroso-propane[57)]. Under somewhat changed conditions corresponding aminyloxides with more than one molecule styrene inserted could be detected. Treatment of styrene and substituted styrenes with nitrogen dioxide in the presence of nitrosobenzene gives analogous γ-nitroalkyl aminyloxides[58)].

Irradiation of 2-methyl-2-nitrosopropane in presence of maleic anhydride does not give di-tert-butylaminyloxide *2b* but a radical to which structure *35* was ascribed as was shown by ESR spectroscopy. Apparently the photolytically generated tert-butyl radical preferentially attacks maleic anhydride forming radical *34* from which spin adduct *35* is derived[25a)]. Addition of iodine to nitroso compound *36* yields aminyloxide *37* by ring closure[59)].

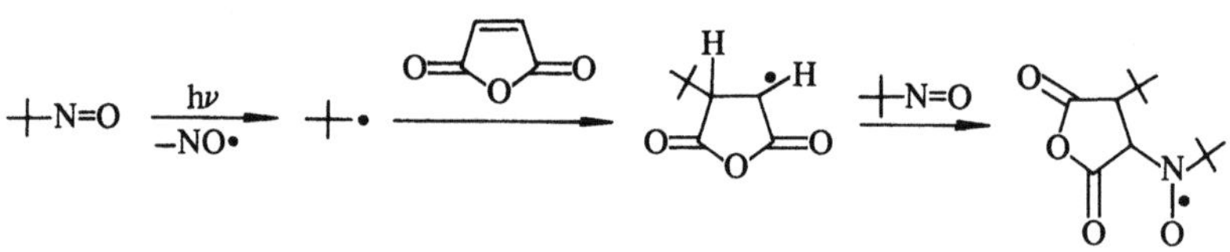

34 *35*

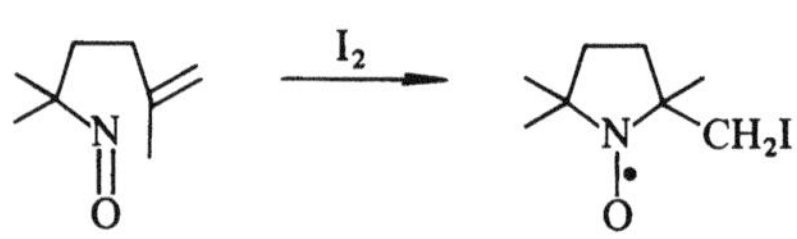

36 *37*

Heating crotonylperoxide in presence of [D_9]-2-methyl-2-nitrosopropane *39* a two-fold spin trap process must be assumed. The generated 1-propenyl radical *38* adds to nitroso compound *39* forming vinylaminyloxide *40*. The latter one, reactive at vinyl carbon (see below, p. 91), is trapped by excess *39* giving aminyloxide *41*[60)].

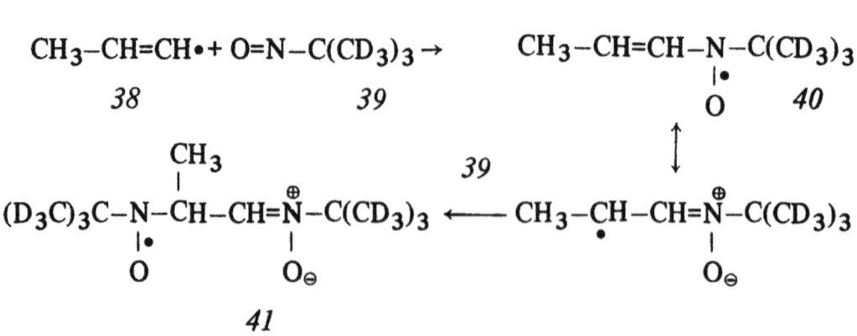

Nitrosoamines can also operate as spin traps. A very impressive example is the appearance of aminyloxide *43* in decomposition of acylarylnitrosoamines[61]. In this reaction phenyl radicals are generated which are trapped by starting compound *42* in a side reaction. The structure of *43* was confirmed by its independent formation from aminyl radical *44* being trapped by nitrosobenzene[62].

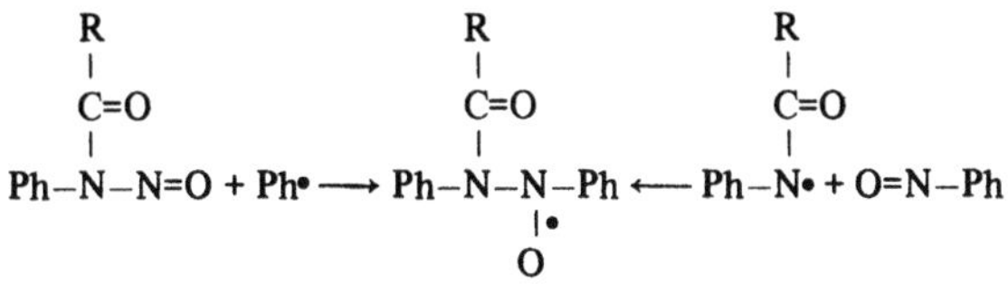

42 *43* *44*

The generally rather unstable alkyl-[63] and arylaminyloxides[64] *2* R = alkyl or aryl, R′ = H are obtained by reduction of nitroso compounds. For instance 1-hydroxy-1-methylethyl radical generated from isopropanol by hydrogen abstraction does not give a spin adduct with 2-methyl-2-nitrosopropane as do radicals derived from primary alcohols but gives tert-butyl-aminyloxide *2* R=$(CH_3)_3C$, R′=H by hydrogen transfer[48d]. Reaction of nitroso compounds and dithionite does not lead to radicals like *45* as supposed by Russell[64]. In fact $\cdot SO_2^{\ominus}$ radical anion is trapped by nitroso compound forming *46*[65]. Addition of Grignard compounds to nitroso compounds also gives aminyloxides[66]. In analogy to reaction with ketones proceeding either by a polar mechanism or by a single-electron-transfer mechanism[67] principally formation of aminyloxides is possible in two ways as represented in scheme (p. 72). Either the radical formed from Grignard compound may be trapped by nitroso compound or an anionic addition takes place, the resulting adduct being oxidized by excess nitroso compound or oxygen. But more frequently aminyloxides have been prepared from Grignard compounds and the more readily available nitro compounds (see below).

Vinylaminyloxides *48* could be obtained by treatment of 2-methyl-2-nitrosopropane with lithium vinyl compound *47*[68]. In contrast to the intermediate vinylaminyloxide *40*, *48* is prevented from secondary reactions apparently by the small spin density at the β-carbon atom, which is caused by the strong twisting of the N–C bond.

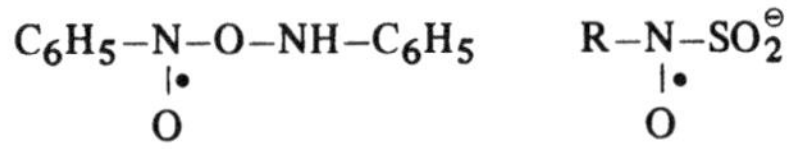

45 *46*

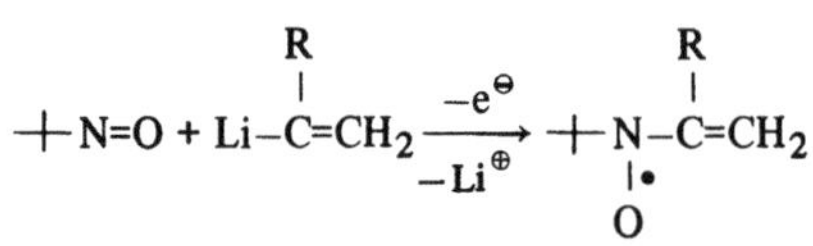

47 *48*

Reaction of nitriles *49* with nitrosobenzene and potassium tert-butylate under exclusion of oxygen gives vinylaminyloxides *51*, the intermediate nitrones *50* being not isolable[69]. Apparently the nitrosobenzene operates as oxidizing agent to some extent.

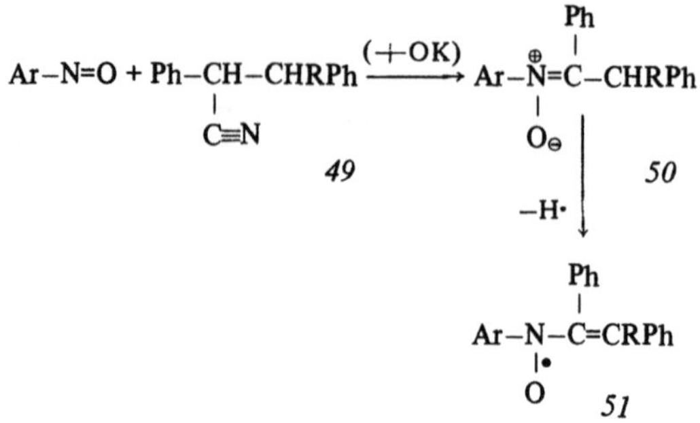

49 50 51

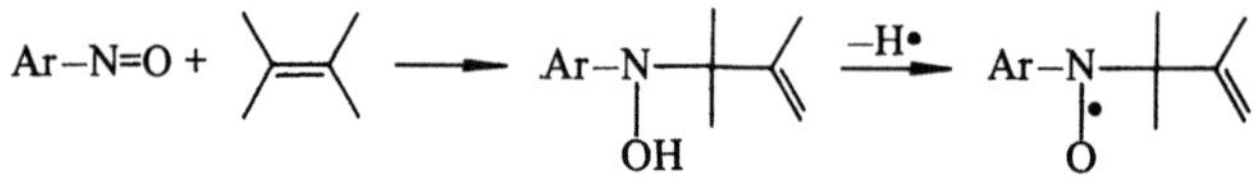

52 53 54

This is also true for reaction of nitrosobenzene[70] and pentafluoronitrosobenzene[71] respectively with 2.3-dimethyl-Δ^2-butene *52* or similar alkenes. This reaction may be interpreted as en-reaction leading to hydroxylamine *53* which is then dehydrogenated to *54*. Formation of aminyloxides *56* from nitrosobenzene and imines *55*[72, 73] is explained analogously as reaction of the tautomeric enamine form[72].

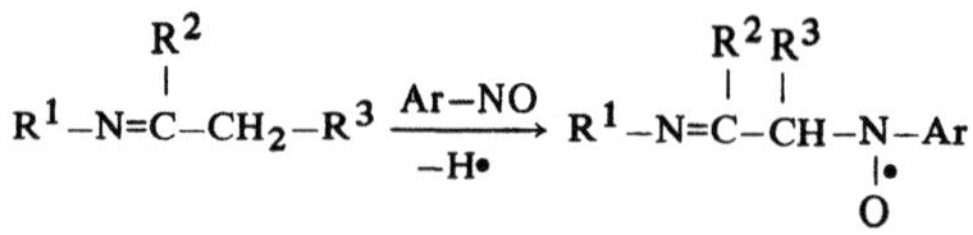

55 56

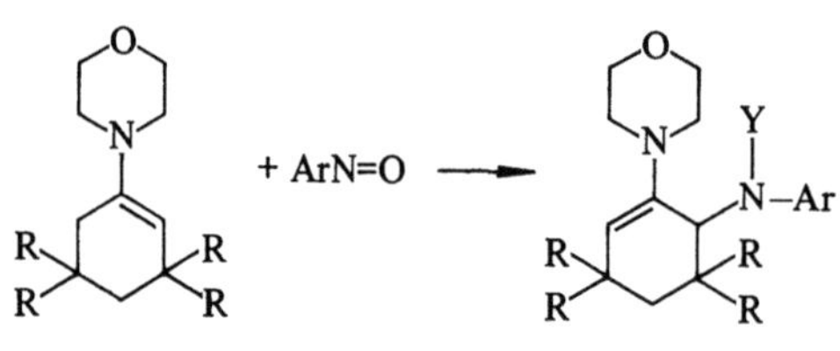

58 : Y = OH
59 : Y = O•

57

R=H or CH_3

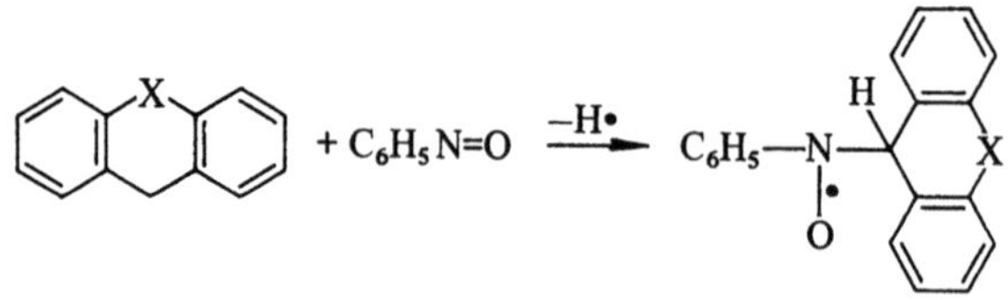

60 61

X = >CH₂, >O, >S, >C=O

Enamines *57* react very easily with nitrosobenzene giving hydroxylamines *58*[74]. Immediately after the mixing of the components the corresponding aminyloxides *59* are detectable in the reaction mixture by ESR spectroscopy[73].

Treatment of dihydroanthracenes and heteroanalogs *60* respectively with nitrosobenzene forms aminyloxides *61* without addition of an oxidizing agent[47c].

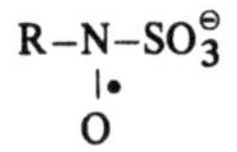

62 R = C_6H_5
63 R = CF_3

Finally mixing of aliphatic or aromatic nitroso compounds with benzene sulfinic acid leads immediately to formation of benzene-sulfonylaminyloxides[75]. Aminyloxide anion *62* was detected when nitrosobenzene was allowed to react with aqueous hydrogensulfite in presence of lead dioxide[76]. With nitroso trifluoromethane radical *63* was formed. With tetraphenylarsonium cation this radical anion even formed a solid, paramagnetic precipitate[77].

4. Aminyloxides from Nitro Compounds

Different types of aminyloxides have been obtained from nitro compounds. Attack of radicals at the oxygen atom of the nitro group gives alkoxyaminyloxides *64*. On the other hand the nitro group may be reduced to a nitroso group during the course of reaction, typical spin adducts *2* of nitroso compounds derived from alkyl radicals R'· being observed. Finally, attack of nucleophiles at the nitrogen atom of the nitro group leads to adduct *65* which decomposes to aminyloxide *2* in the next step.

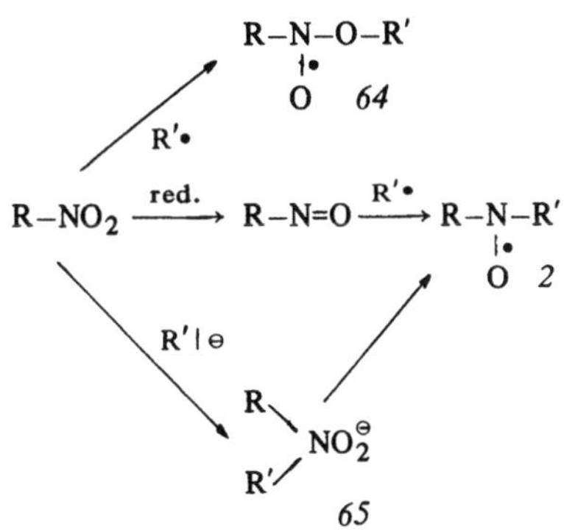

Usually alkoxyaminyloxides are formed only in reactions with electron-rich alkyl radicals, which are derived for instance from ethers or alcohols by hydrogen abstraction. Hence it is supposed that a charge transfer in the transition state or in an intermediate *68* plays an important role. During photolysis of aromatic nitro compounds in ethers or alcohols alkoxyaminyloxides *69a* could be detected[78]. The photochemically excited nitro compound abstracts a hydrogen atom from solvent. Alkyl radicals *67* so formed react with nitro compound giving *69a*. The importance of charge transfer between alkyl radical and nitro compound is shown by the fact that nitro compounds with strong electron-donating substituents do not react[78a]. The hydroxy-aminyloxide *66a* usually could not be detected as it apparently suffers easy secondary reactions to diamagnetic products. Only during photolysis of 2,3,5,6-tetrachloro- and 2,4,6-trichloronitrobenzene is an ESR signal obtained that is ascribed to radical *66a*[78b].

In many photoreactions of aromatic nitro compounds secondary radicals are observed[78b]. Whereas primary radicals *69a* are detectable only during irradiation secondary radicals are formed to a larger extent frequently only after irradiation has ceased. Unsubstituted nitrobenzene gives the secondary radical phenylaminyloxide. More information with respect to secondary radicals comes from the investigations of polychloronitrobenzenes with substituted *ortho* positions. In these cases radicals of type *73* and *74* respectively are obtained, which are spin adducts of the corresponding nitroso compound. The nitroso compound may be formed either by direct reduction of nitro compound *via* radical *66a* or by decomposition of primary radical *69a* which occurs very easily (see below, α-scission). The alkoxy radical *70*, being generated in this way, fragments to ester *72* and an alkyl radical *71*, the latter being trapped giving *73*. The other aminyloxide *74* is formed by trapping of radical *67*.

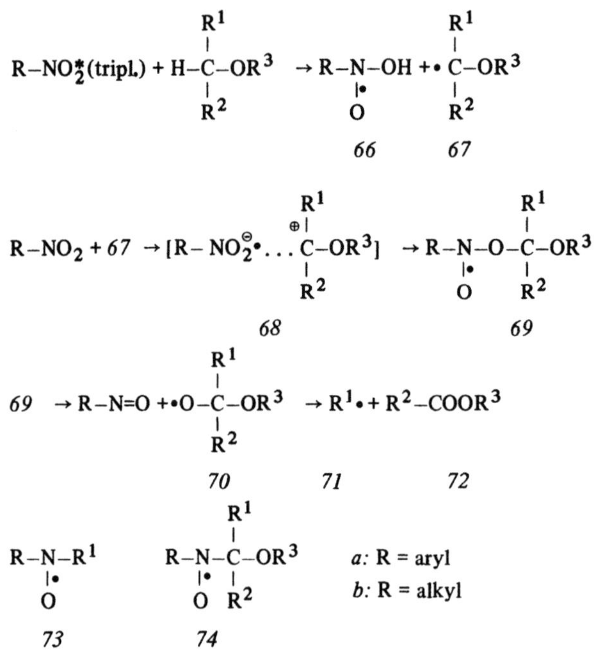

In reaction of aliphatic nitro compounds with alkyl radicals *67* generated from ethers or alcohols aminyloxides *69b* could be detected[79]. Moreover dialkylaminyloxides *74b* are formed, *67* being trapped from the corresponding nitroso compound. Reduction of nitro compound to nitroso compound probably occurs by electron transfer from alkyl radical *67* to nitro compound, subsequent dissociation of the resulting complex *68b* giving nitro anion radical which finally disproportionates.

In contrast to the results of Sutcliffe, Chachaty[80] could not observe alkoxyaminyloxides *69a* but only secondary radicals *74a* (R^3 = H), irradiating unsubstituted nitrobenzene in different primary alcohols. Irradiation of aliphatic nitro compounds gave the same results[81].

Recently A. Berndt[82a] obtained aminyloxides *2* R = alkyl or aryl, R′ = tert-butyl by reaction of photochemically generated tert-butyl radicals with nitro aliphatics, nitro aromatics and suitably substituted nitroalkenes respectively. In this case it is also supposed that tert-butoxyaminyloxides *64* R′ = tert-butyl are formed first as was established by low temperature studies[82b]. In some other cases, especially when the nitro group is sterically hindered, the corresponding tert-butoxyaminyl radicals have been obtained. Sometimes rearrangement of these radicals to aminyloxides *2* R′ = tert-butyl is observed. Only recently silyl radicals and stannyl radicals have been trapped by nitro compounds yielding the corresponding aminyloxides *64* (R′ = SiR_3[82c–e] or R′ = SnR_3[82b]).

Another interesting example of formation of aminyloxides is the thermal decomposition of benzoylperoxide in benzene in presence of nitrobenzene. Reduction of nitrobenzene to nitrosobenzene to some extent and subsequent spin-trapping of phenyl radical gives diphenylaminyloxide[83].

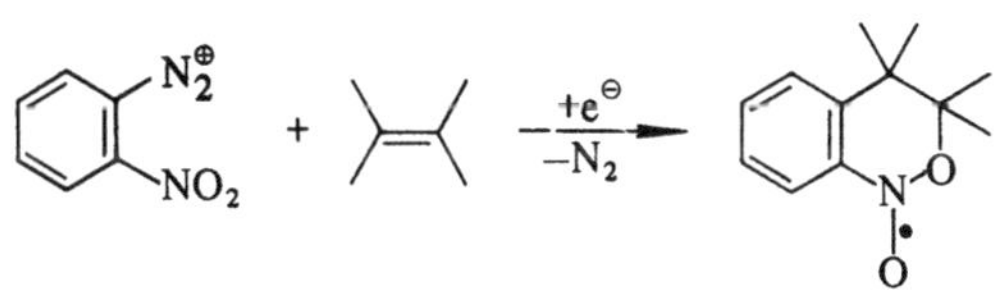

75 *76*

Reduction of o-nitrodiazoniumsalt *75* in presence of 2.3-dimethyl-Δ^2-butene leads to aminyloxide *76*. The first formed o-nitrophenyl radical adds to the olefin, *76* being formed by a subsequent intramolecular spin-trapping process[84].

The anion radical of 2-methyl-2-nitropropane being formed by reduction with alkali metal or electrolysis fragmentates to tert-butyl radicals and nitrite ions. Thus formation of di-tert-butylaminyloxide is initiated[85]. Treatment of 2-methyl-2-nitropropane with sodium phenyl gives tert-butyl-phenylaminyloxide the reaction proceeding through an intermediate hydroxylamine oxide salt *65* which is even isolable[86]. The frequently used method for synthesizing aminyloxides, the reaction of Grignard compounds, with aliphatic or aromatic nitro compounds, is also thought to proceed through an analogous intermediate *65*[87].

5. Aminyloxides from Nitrones

In the same way as nitroso compounds, nitrones are also used as spin traps. Reactive radicals are added to the carbon atom of the nitrone group and thereby converted to relatively stable aminyloxides[38, 88]. But there is also the possibility of adding a nucleophile to the nitrone in a non-radical way giving adduct *77* which is oxidized to aminyloxide in a second step. In this way aminyloxides containing a variety of different groups R have been detected. Alkyl- and aryl groups[89] are trapped mainly, but aminyloxides containing the heteroatoms O[90] and N[91] in the β-position have been obtained in this way also. Mainly N-benzyliden-tert-butylamin-N-oxid *81 c*[88],

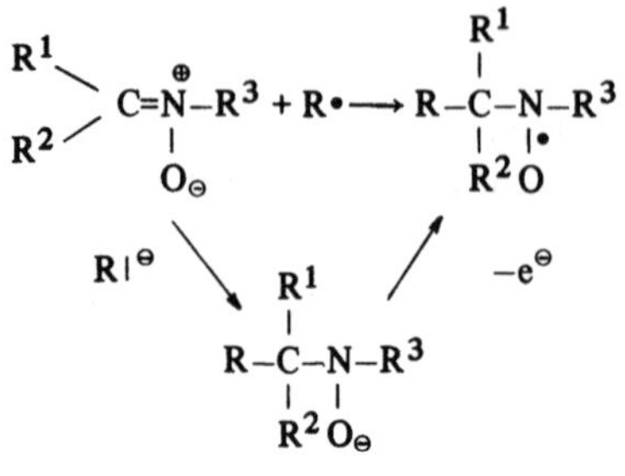

77

N-methylen-tert-butylamin-N-oxid *79*[91a] and 5,5-dimethyl-Δ^1-pyrrolin-N-oxid *78*[92] have been employed as spin traps, but a variety of other nitrones is also suitable[93]. While ESR data from adducts to *81c* give relatively little information on fine structure of the radicals[88], in the spectra of adducts from *79* and *78* more such information is recognizable[92c, d, 94].

By ESR spectroscopy Janzen measured the rate of formation of aminyloxides from benzoyloxy radicals and N-aryliden-tert-butylamin-N-oxides *81* and so determined the rate of thermal decomposition of benzoylperoxide[90c]. The value estimated in this manner is in accordance with those determined by other methods. In the same way the rate constants for spin trapping of tert-butoxy radicals by different spin traps were determined[90b]. The relative values listed in the following table show how effectively the spin traps operate.

Relative rates for trapping of tert-butoxy radicals

			$k_{rel.}$
78	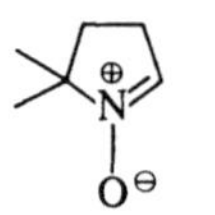		1
79	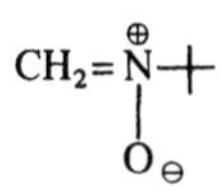		0.5–1
80	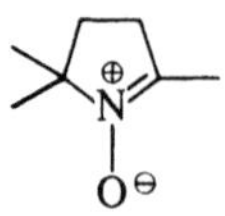		0.42
81a	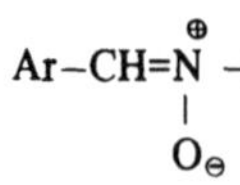	p–NO_2	0.018
81b		p–Cl	0.013
81c		–	0.011
81d		p–OCH_3	0.011
81e		p–CH_3	0.0068
	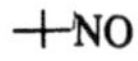		0.0030

Electrochemically generated radicals, as for instance phenyl radicals formed by reduction of phenyldiazonium tetrafluoroborate, may be trapped by *81 c* and detected by the ESR spectrum of the corresponding aminyloxide[95].

Nitrone *82* may be used to distinguish carbon radicals and oxygen radicals. With alkyl- and aryl radicals addition to the nitrone group with formation of aminyloxides

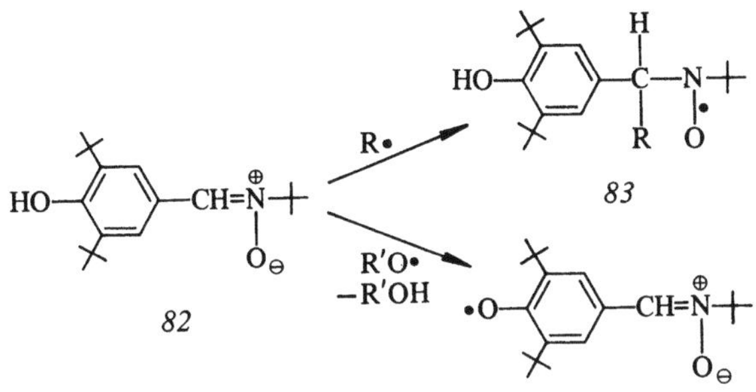

82 *83* *84*

R = CH_3, $(CH_3)_2COH$, $(CH_3)_2CCN$, C_6H_5

R'O = CH_3COO, C_6H_5COO

83 occurs, whereas with oxygen radicals by hydrogen abstraction *84* is formed[96]. A 50 : 50 mixture of benzyliden-tert-butylamin-N-oxide (*81 c*) and 2,4,6-tri-tert-butylphenol operates in the same way[97].

Formation of aminyloxides *86* from primary or secondary amines and nitrone *79* requires lead dioxide as oxidizing reagent. Adduct *85*, being formed only to a small extent in equilibrium with starting compounds, is oxidized giving *86*[98]. Radicals *86* (R' = H) derived from primary amines may be oxidized further (see below p. 96). 4,5,5-Trimethyl-Δ^1-pyrrolin-N-oxide also has been employed in this oxidative coupling reaction. Reaction of nitrone *79* with tert-butylhydroxylamine and lead dioxide gives mainly O-adduct *87*, the N-adduct *88* could be detected only under specific reaction conditions.

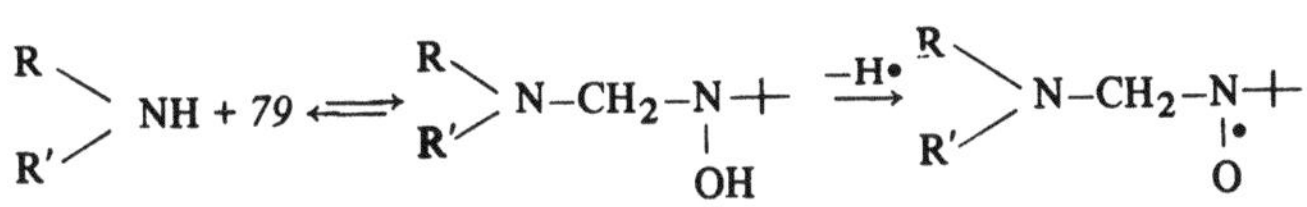

85 *86*

R' = H or alkyl

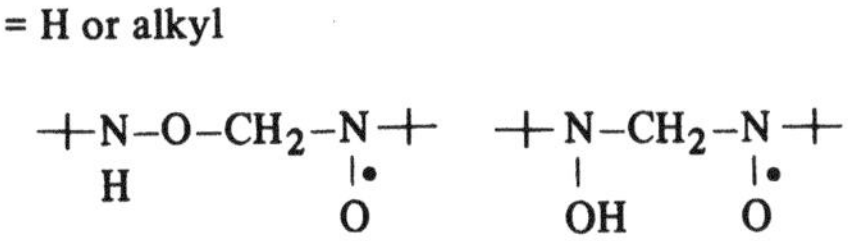

87 *88*

Acylaminyloxides *90* were obtained in photolysis of *81 c* and *89* respectively[99]. The mechanism of formation of these radicals is not yet known. It is suggested that formation of acylaminyloxides *90* occurs from photochemically excited oxaziridines since *90a* is also formed by irradiation of oxaziridine *91*. On the other hand heating of *81 c* in presence of oxygen gives acylaminyloxide *90a* and an aminyloxide for which structure *92a* is supposed. With thorough exclusion of oxygen only di-tert-butylaminyloxide and a radical, probably having structure *92b* could be detected[100].

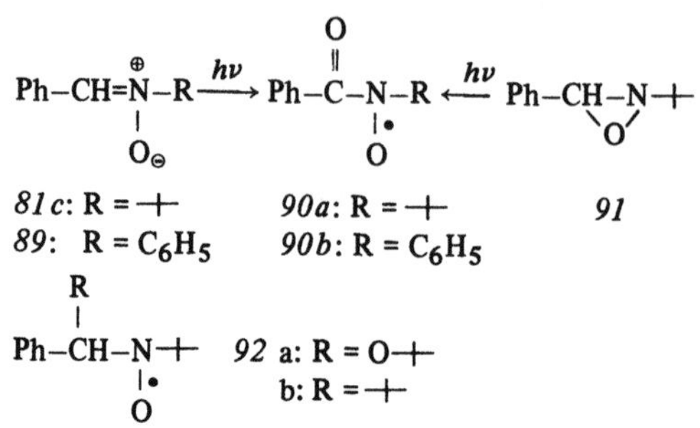

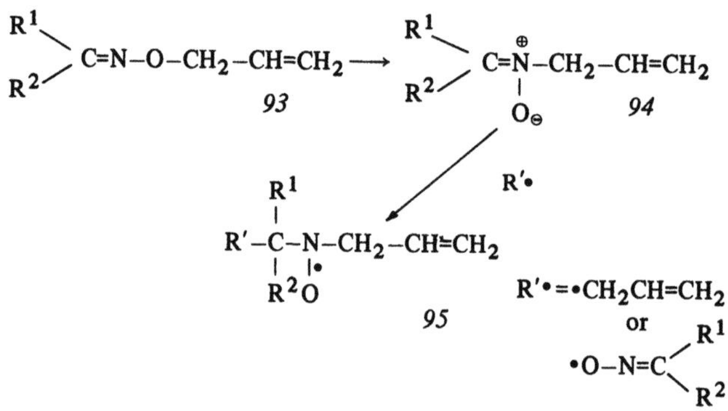

The Meisenheimer rearrangement of oxime-O-allylethers *93* giving nitrones *94* is accompanied by formation of aminyloxides[101]. It is assumed that one of the radical fragments arising from *93* is trapped by *94* giving radical *95* with two alternative possibilities for R′.

6. Aminyloxides from Oximes

In the oxidation of oximes with lead-tetraacetate, aminyloxides are observed occasionally besides iminyloxide radicals (iminoxyles)[102]. Thus in oxidation of bicyclic oximes three types of aminyloxides could be detected[102a]. In one case nitroso compound *96* could be isolated. The well-known conversion of similar nitroso compounds to corresponding aminyloxides[103] together with ESR data makes structure *97* probable for the first type of aminyloxides. For both of the other types of aminyloxides the structural features *98* and *99* could be recognized but their exact structure could not be determined. Surprisingly action of hydroxy radicals on aliphatic oximes does not give iminyloxides but β-hydroxy aminyloxides *101*[104]. They are formed by addition of hydroxy radicals to the carbon atom of the oxime group yielding *100* and subsequent migration of hydrogen. Amino radicals and hydroxymethyl radicals react in an analogous manner but methyl radicals do not react[104a].

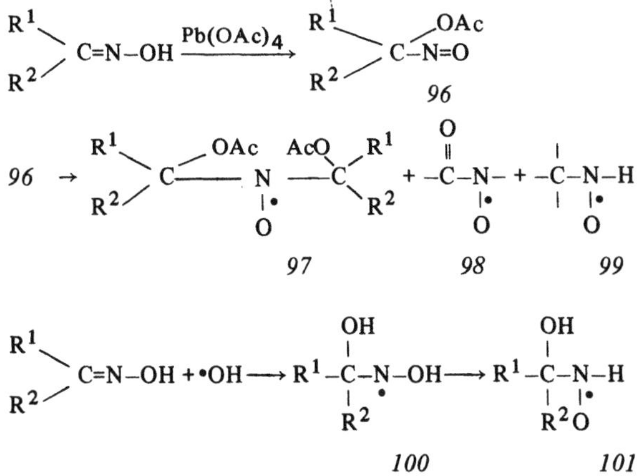

IV. Reactions of Aminyloxides

In contrast to more simple principal reactions of aminyloxides which have been reviewed previously[3, 4b, 5], the reactions now being reviewed are frequently more complex. Thus classification of reaction types is sometimes very difficult and overlapping of some Chapters could not be avoided.

1. Reactions without Direct Involvement of Aminyloxide Group

Reactions of this type have been investigated mainly by Rozantsev and co-workers[4b]. Tetramethylpiperidonyloxid *102* was selected as an example for demonstrating the possibilities of reactions without involvement of radical function (scheme), all reactions occurring with commonly employed reagents. Thus typical carbonyl group reactions with N-nucleophiles, CH-acid compounds and Grignard compounds respectively, may be performed with *102*, while reaction with formaldehyde[105] gives rise to formation of condensation products. Likewise the hydroxy group of reduction product *103* may be converted into an ester function. The variety of reactions of *102* is actually considerably greater, in nearly all reactions the radical center remains unchanged. Because of these unusual properties *102* and some of its derivatives were employed as spin labels, that means the radical is connected with another molecule by the functional group introducing thereby a radical label into this molecule[106]. Other frequently used spin labels are *104–107*, also containing reactive functional groups suitable for connecting them with other molecules.

In all these radicals the aminyloxide group is sterically hindered by two bulky tertiary alkyl groups. Thus it was thought that the high stability of the radical center is caused by steric effects. But it is shown by the results of Ullman[108] concerning the reactions of the cyclic amidinyl-N.N′-dioxide *108*, the tetramethylimidazolinyl-1.3-dioxide, that steric factors are not the only reason for the remarkable stability

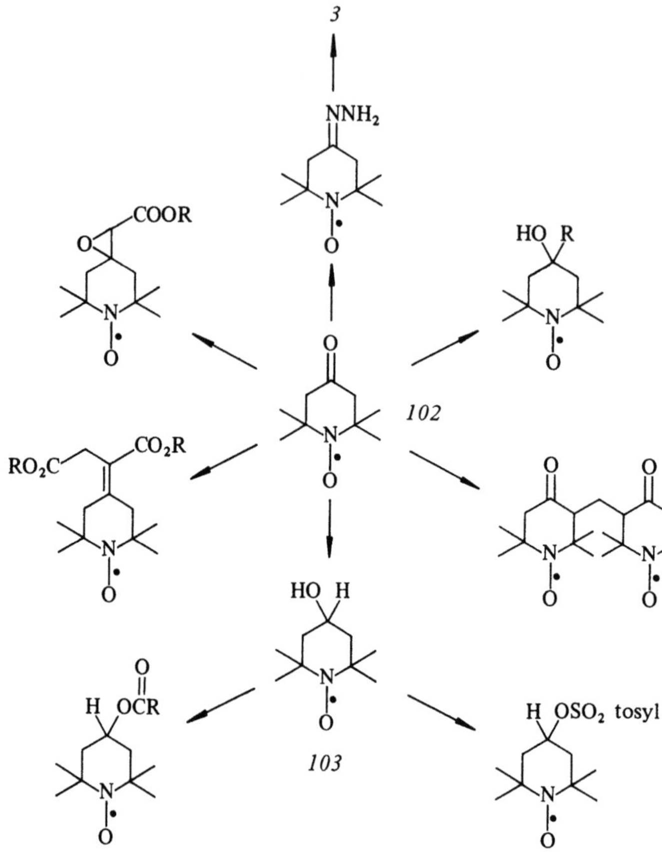

of these radicals. Apparently this stability originates to a considerable extent from the general stability of aminyloxide function arising from its low energy content. Thus *108* suffers a variety of reactions with retention of aminyloxide function. *108* forms anion *109* under basic conditions, while under acid conditions the hydrogen atom may be exchanged by deuterium[109]. With halogen, *108* forms halogenated radicals *110*. *110* (X = J) is converted to the conjugated biradical *111* in the presence of copper[110]. In *110* (X = Br) a nucleophilic displacement of bromine by cyanide ion is possible, yielding *112*, while by action of hydroxide ion anion radical *114* is formed whereas treatment with ammonia gives radical *113*[111]. Though these reactions all occur in the immediate vicinity of the radical center, the latter is not affected.

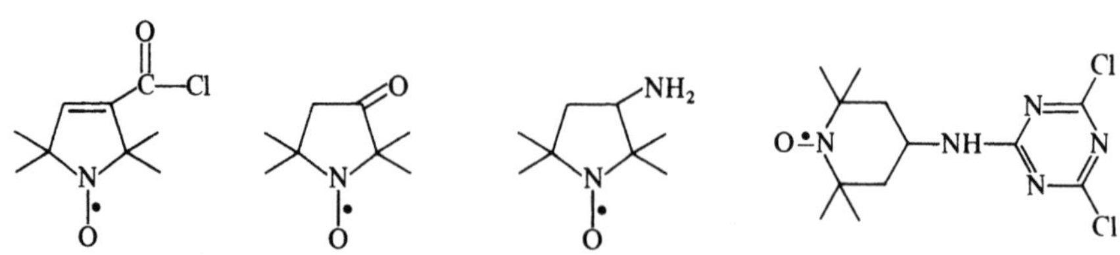

104 *105* *106* *107*[107]

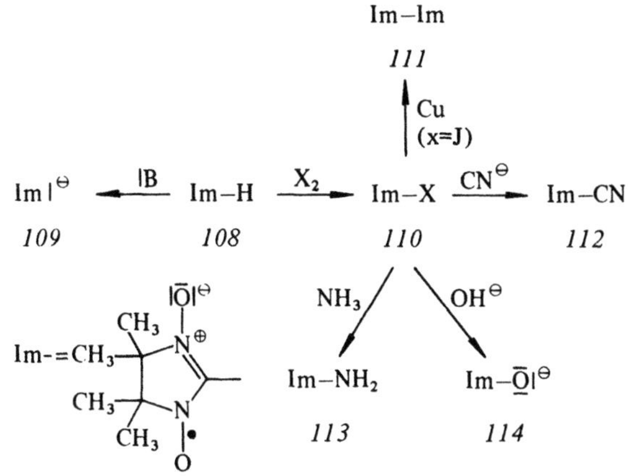

Reactions may also occur in the side chain of such radicals. For instance *115* may be deprotonated to α-sulfonyl-carbanion radical *116*[112]. In the radicals *117* the halogen atom may be substituted by O- or N-nucleophiles, the order of reactivity being Cl < Br < J as expected[113]. Reaction of *117* with benzene sulfinic acid leads to *115*[112]. Reactions may also be performed with biradicals, without effecting the radical character. *119* adds bromine giving biradical *120*, from which *119* may be regenerated by treatment with hydroxide ions[113].

$$\underset{115}{Im-CH_2-SO_2C_6H_5} \xrightarrow{|B} \underset{116}{Im-\overset{-\ominus}{C}H-SO_2C_6H_5}$$

$$\underset{117}{Im-CH_2-X} \xrightarrow{Nu} \underset{118}{Im-CH_2-Nu} \qquad X = Cl, Br, J$$

$$\underset{119}{Im-CH=CH-Im} \underset{(OH^{\ominus})}{\overset{Br_2}{\rightleftharpoons}} \underset{120}{Im-CHBr-CHBr-Im}$$

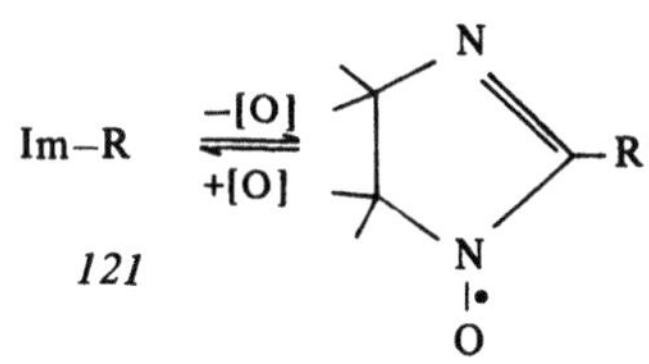

121

122

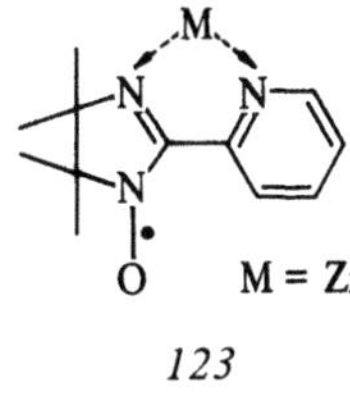

M = Zn,Cd,Hg

123

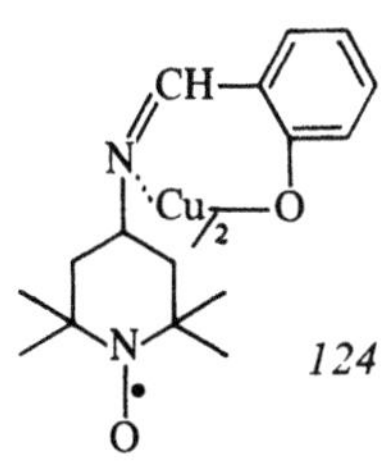

124

By heating with triphenylphosphine or nitrous acid amidinyl-N.N'-dioxides *121* may be de-oxygenated yielding the amidinyl-N-oxides *122*[114]. On the other hand radicals *122* may be reconverted to *121* by oxidation with hydrogen peroxide in the presence of phosphotungstic acid. Like *121*, amidinyl-N-oxide compounds *122* give reactions with preservation of radical function and are also employed as spin labels[113].

Another type of reaction without involvement of radical function is the complexation of metal ions with aminyloxides containing chelate-forming groups. B. E. Wagner[115] studied complexes of type *123*. ESR studies showed a considerable coupling of the unpaired electron with metal nuclei (M = Cd, Hg). Chelate-complex *124* is paramagnetic as expected[116].

The above-mentioned reactions start with very stable radicals which can be isolated in most cases. On the other hand reactions of less stable aminyloxides which could not be isolated have been studied by ESR spectroscopy. One example of these reactions is nucleophilic substitution at chlorocarbonyl-aminyloxide *125a*[60, 117] giving radicals *125b–e*. These reactions all take place directly in the environment

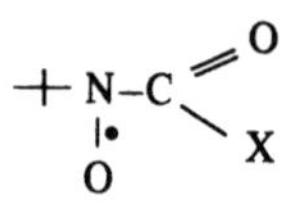

125

a: X = Cl *b:* X = NRR' *c:* X = NHR *d:* X = N(OH)R *e:* X = OR

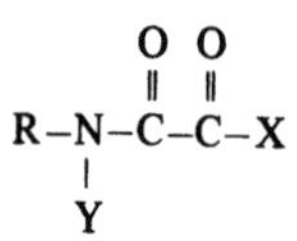

126: Y = O• X = Cl
127a: Y = OH X = Cl
b: Y = OH X = NHR'

of the radical center without changing it. But it may not be excluded that in any case the aminyloxide group affects the course of reaction.

A directing effect of the radical center is found in reaction of isatinyl-N-oxide with primary aliphatic amines, giving 2-alkylimino-isatinyl-N-oxides[118]. In this case the reactivity of carbonyl group neighbored to aminyloxide group is enhanced, as is shown by the fact that the comparable N-hydroxy-isatin as well as isatin itself reacts at the carbonyl group at 3-position[119]. Also reaction of N-(2-chlor-oxalyl)-aminyloxides *126* with amines leading to carbamoylaminyloxides with loss of carbon monoxide proceeds contrary to those of the reduced parent compounds *127a* giving *127b* as expected[23a].

2. Reactions Involving Aminyloxide Function

The principal types of reactions at aminyloxide group are shown in the following scheme. More complex reactions may also be derived from these principle types of reactions.

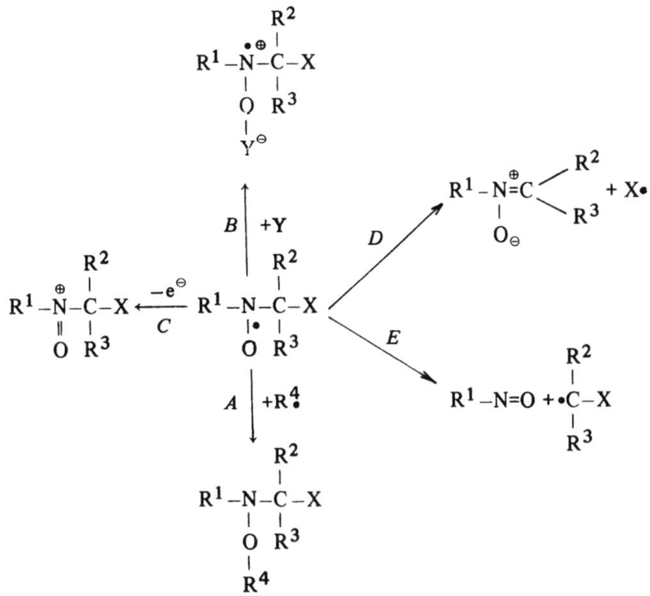

A: addition of radicals (R^4 = H: reduction)
B: addition of Lewis acids ($Y = H^+$: protonation)
C: oxidation
D: β-scission
E: α-scission

a) Addition of Radicals to Oxygen Atom of Aminyloxide Group

Typical radical reactions are additions of other radicals forming a diamagnetic product. Aminyloxides mainly react at the oxygen atom.

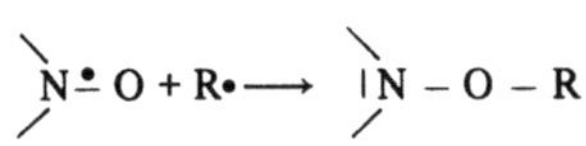

The first investigations of such reactions were undertaken by Wieland and coworkers studying reactions of diarylaminyloxides with triphenylmethyl or nitrogendioxide[120)]. But in only a few cases could the simple addition products be isolated, for mostly complex subsequent reactions took place. Many examples of trapping radicals by aminyloxides are recorded by Rozantsev[4b)] or Forrester[5)]. In this connection inhibition of radical reactions by aminyloxides was discussed[4b)].

Reaction with Grignard compounds may also be interpreted as radical addition[121)]. In this reaction *3* is partly reduced forming salt *128*, partly it traps radical R yielding *129*.

The radicals formed in photolysis of dibenzylketone are trapped by tetramethylpiperidyl-N-oxide *3* giving the corresponding N-benzyloxypiperidine and N–(2-phenyl)-acetoxypiperidine as expected[122)]. CIDNP-studies were performed with acetylbenzoylperoxide being decomposed in the presence of *3*[123)]. The appearance of the

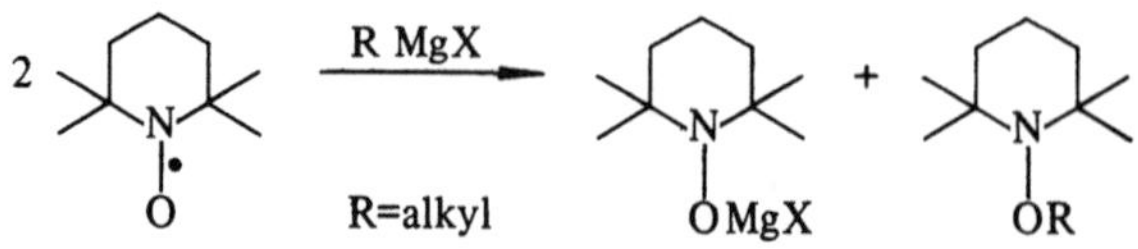

3 128 129

corresponding N-methoxypiperidine as the single trap-product indicates that non-aromatic aminyloxides like *3* trap only alkyl radicals but not phenyl radicals[124].

A special case of generation of radicals being trapped by aminyloxides is the abstraction of hydrogen from solvent. Photochemically excited 4-hydroxy-tetramethyl-piperidyl-N-oxide (*103*) abstracts hydrogen from solvent RH, the resulting R· being trapped by a further molecule of *103*. Thus the corresponding 1,4-dihydroxy-piperidine and 1-alkoxy-4-hydroxy-piperidine are formed in equal yield[125]. The relative rate of photolysis increases with increasing stability of radical intermediate R·. (RH = cumene, toluene, cyclohexane: $k_{rel} = 420 : 1, 1 : 1.$)

An interesting example of hydrogen abstraction by aminyloxide function is the slow decomposition of the relatively stable aminyloxide *102*[126]. In this case a second molecule of *102* acts as hydrogen donor giving diradical intermediate *130* which is trapped by a further radical *102* giving *132* *via* *131* at room temperature. With heating, *130* decomposes, yielding phorone *133* (α-scission).

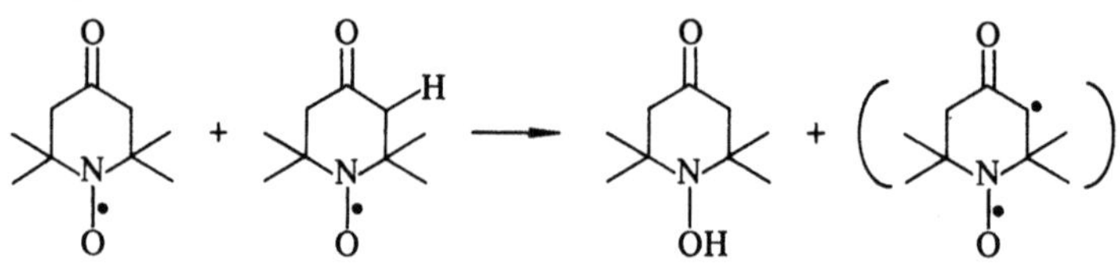

102 130

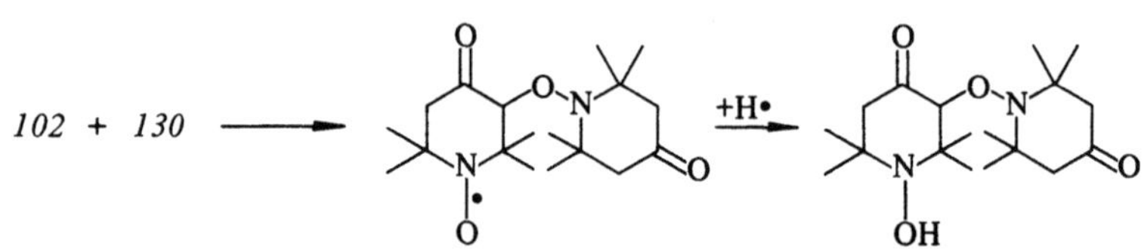

131 132

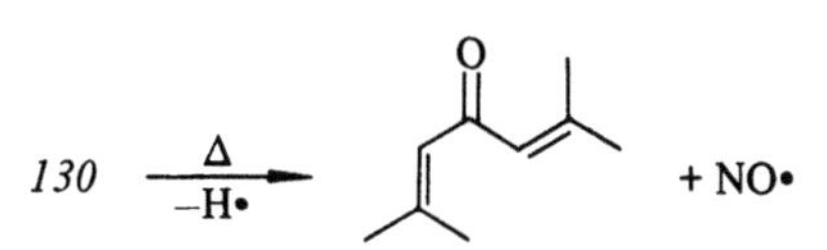

133

Abstraction of hydrogen may also occur from a hydroxylamine being oxidized to a new aminyloxide in this way[108, 127].

$$\underset{\substack{|\bullet \\ O}}{R^1{-}N{-}R^2} + \underset{\substack{| \\ OH}}{R^3{-}N{-}R^4} \rightleftharpoons \underset{\substack{| \\ OH}}{R^1{-}N{-}R^2} + \underset{\substack{|\bullet \\ O}}{R^3{-}N{-}R^4}$$

134 *135–H* *134–H* *135*

The position of equilibrium between two different aminyloxides *134* and *135* is determined by their relative stability with respect to the corresponding hydroxylamines *134–H* and *135–H*. Thus the OH-bond-dissociation energies may be used as a measure for the stability of corresponding aminyloxides. For hydroxylamines *3–H, 102–H, 136–H* and *137–H* the following values were determined. From these values it is apparent, that the equilibrium *136* + *3–H* ⇌ *136–H* + *3* is strongly shifted to the right[22].

	D kcal/Mol
3–H	71.8
102–H	69.6
136–H	77.0
137–H	76.2

136

137

A special situation develops if aminyloxide group and hydroxylamine group are in the same molecule. For radical *138* it is shown by labeling one nitrogen atom with ^{15}N-isotope that both nitrogen atoms couple with the unpaired electron indicating an equilibrium between *138A* and *138B*[128]. For radical *139* it was shown by a tem-

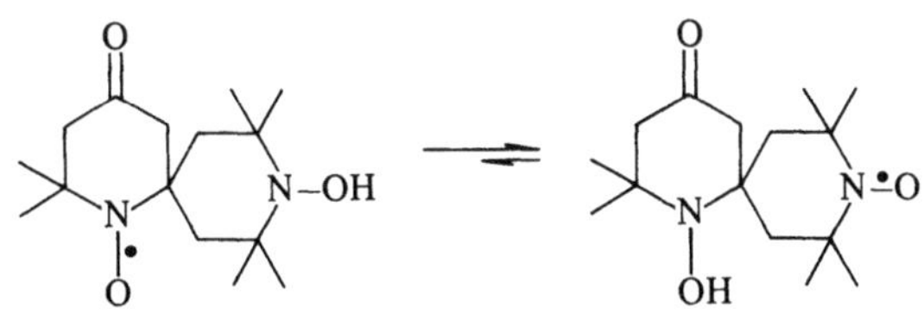

138A *138B*

139A *139B*

perature-dependent ESR study that an intramolecular hydrogen exchange occurs, the activation energy being 4.5 kcal/Mol[b] for this process[117c].

With respect to other aminyloxides bis-trifluoromethylaminyloxide takes an exceptional position. As was already mentioned, the inductive effect of the trifluoromethyl groups causes a high spin density at oxygen atom (p. 68). On the one hand the purple gas is unreactive towards water and benzene[20] but on the other hand it reacts violently with a variety of reagents[129–135]. Hydrogen abstraction and subsequent radical addition occur with hydrocarbons[130]. Even the hydrogen of an aldehyde group may be abstracted by bis-trifluoromethyl-aminyloxide[131]. With

---

[b] In the original paper the value 2.7 kcal/Mol was erroneously given.

olefines[130], tetrafluoroallene[132], acetylenes[133] and isonitriles[134] typical radical reactions arise. The addition of four molecules of bis-trifluoromethyl-aminyloxide to pentafluoropyridine is comparable to photochemical chlorination of this compound[135].

b) Dimerization and Reaction at Conjugated Groups of Aminyloxides

A special case of radical reaction is dimerization. Most typical stable dialkylaminyloxides like *3* and *2b* do not show dimerization. The reason for the failure of dimerization clearly is the unfavorable energy content of any dimer. Assuming a dimerization by O–O-bonding the loss of delocalization energy (p. 67) of two aminyloxide groups apparently cannot be compensated by gain of energy from making a weak O–O-bond. In addition, steric effects further weaken the tendency to dimerization. These steric effects are reduced in bicyclic aminyloxides like *136* and *137*. Furthermore the delocalization energy is also reduced for *136* and *137*, as was clearly shown by the higher bond dissociation energies for corresponding hydroxylamines[22]. Accordingly *136* and *137* dimerize in solution at low temperature reversibly, the structure of the dimer being still unknown[136]. Contrary to typical aminyloxides such as *3* and *2b*, a reversible dimerization in the solid state was found for bis-trifluoromethyl-aminyloxide[20], *140*[137], tetraphenyl-pyrrolyl-N-oxide[138], and the Fremy-salt[139]. X-ray analysis indicates an arrangement *141* for the dimer of *140*[137]. Besides

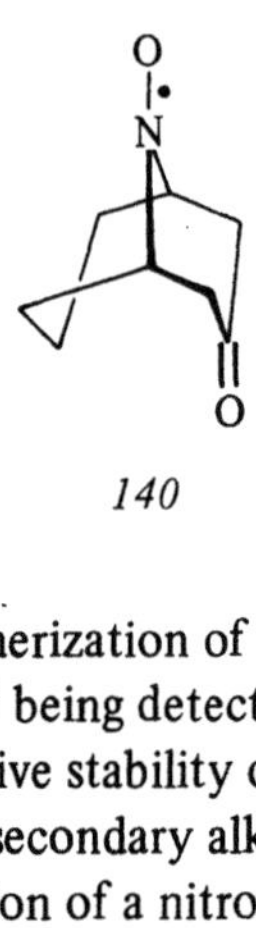

140

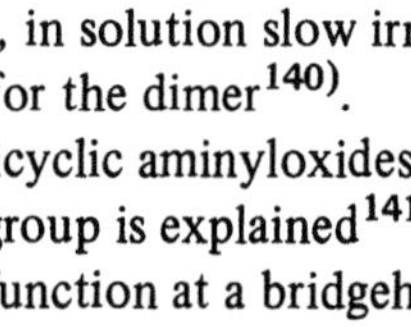

141

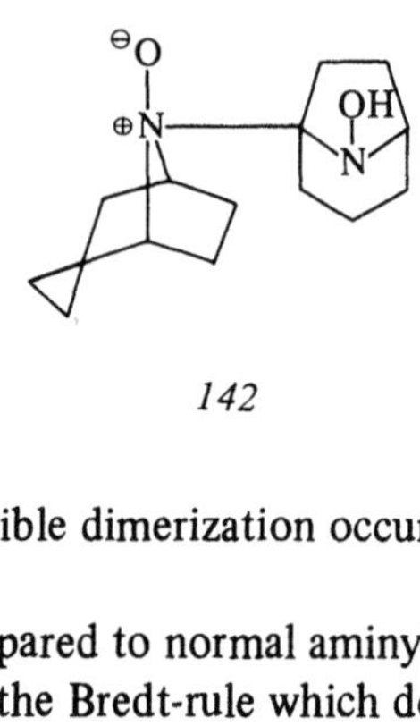

142

reversible dimerization of *136*, in solution slow irreversible dimerization occurs, structure *142* being detected for the dimer[140].

The relative stability of bicyclic aminyloxides compared to normal aminyloxides containing a secondary alkyl group is explained[141] by the Bredt-rule which does not allow formation of a nitrone function at a bridgehead atom. Generally aminyloxides *143* with secondary and primary alkyl groups attached to the radical function disproportionate to nitrones *147* and hydroxylamines *148*[142], thus being far less stable than tertiary alkylaminyloxides. This reaction was kinetically studied by Ingold and coworkers. They found that disproportionation of *143* is a bimolecular reaction and detected a large isotope effect, substituting the β-hydrogen atom by deuterium ($k_H/k_D = 6 - 14$)[144]. Therefore, they postulated an equilibrium with a dimer *144* decomposing slowly *via* *145* to *147* and *148*. Of course the O–O-connection for the dimer could not be proven. Disproportionation *via* a dimeric intermediate *146* also seems reasonable.

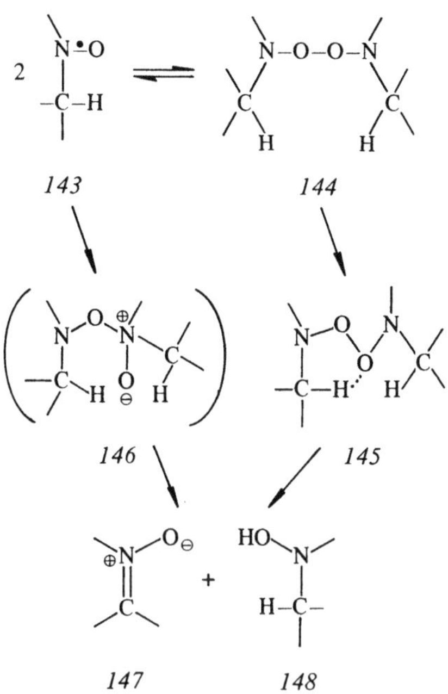

A bimolecular reaction was also found for disproportionation of monoalkyl-aminyloxides *2* R = alkyl R′ = H to corresponding nitroso compounds and hydroxyl-amines[143]. A reversible dimer of N-methyl-N-(2,4,6-tri-tert-butyl)-phenyl-aminyl-oxide, the structure of which is unknown, was detected in solution. The radical decomposes as expected giving the corresponding nitrone and hydroxylamine[145].

Aminyloxides *4* containing a conjugated group offer additional possibilities for dimerization. Generally dimerization may occur through forming of a Y–Y-bond (route A) or an O–Y-bond (route B).

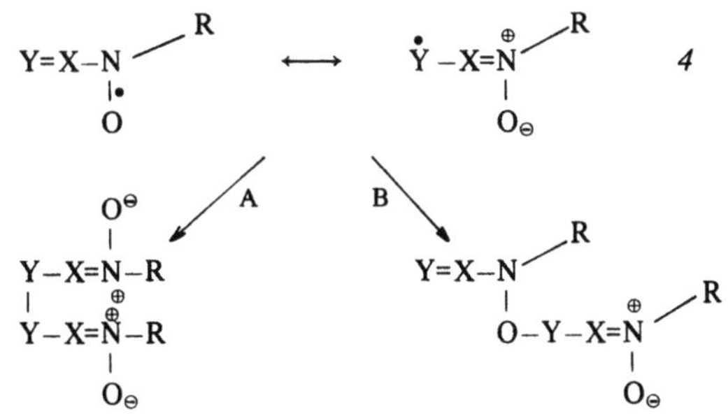

An obvious example for route A is aminyloxide *149*, which could be detected only at enhanced temperature and dimerizes at room temperature with formation of a C–C-bond giving *150*[146]. On the other hand the comparable aminyloxide *151* does not dimerize even at low temperature. Obviously in the latter case, gain of N–N-bonding energy cannot compensate the loss of delocalization energy in dimerization to *152*, whereas gain of C–C-bonding energy in *150* does.

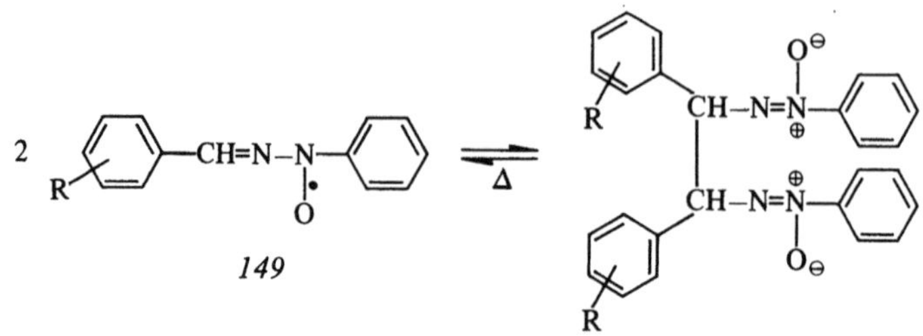

149 150

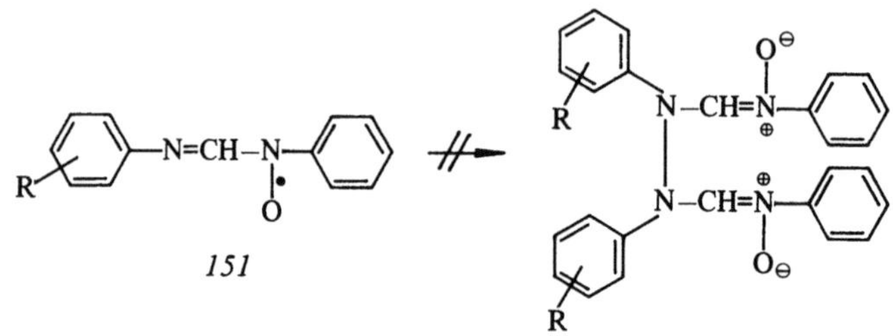

151 152

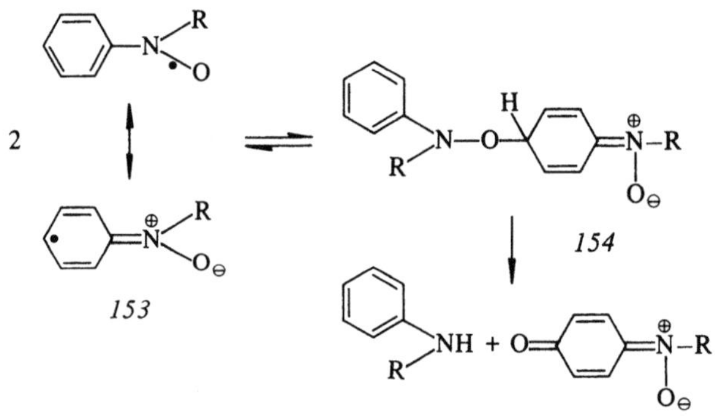

153 154 155 156

Generally aminyloxides *153* with unsubstituted p-position are stable in solution only for a limited time. Decomposition to *155* and *156* occurs by standing in solution for some time or by evaporation to dryness. This reaction is formulated *via* dimer *154*, formed according to route B, which subsequently fragmentates yielding *155* and *156*. The first example studied was the decomposition of diphenylaminyloxide[147].

The mechanism is confirmed by the fact that steric blocking of the para-position by alkyl groups generally stabilizes aryl-substituted aminyloxides[4b, 148]. Two bulky substituents in the m-position give the same effect[148]. But a stabilization is not afforded by halogen substituents at p-position[151]. The corresponding dimer is slowly decomposed to p-benzoquinone-imine-N-oxide (*156*), aminyl radicals and halogen atoms which are converted to arylamines and halogen acid by hydrogen abstraction from solvent. The order of reactivity is p–J < p–F < p–Cl < p–Br. Trapping of the phenyl radical by pentafluoronitrosobenzene yields N-pentafluorophenyl-N-phenyl aminyloxide together with its decomposition products, the corresponding amine and p-benzoquinone-imine[152].

Substitution in the o-position also stabilizes aryl substituted aminyloxides. But in this case the aryl ring is twisted by steric effects reducing delocalization of the unpaired electron into the aryl ring and hence decreasing spin density at the p-position. In this way formation of an intermediate analog to *154* is rendered more difficult. Thus decomposition of tert-butyl-o-tolyl-aminyloxide *157* does not occur *via* a dimer like *154* but by an intermolecular hydrogen abstraction yielding disproportionation products *158* and *159*, the former one giving finally *161* *via* intermediate *160*[149]. Analogous products were found in decomposition of tert-butyl-p-tolyl-aminyloxide[148].

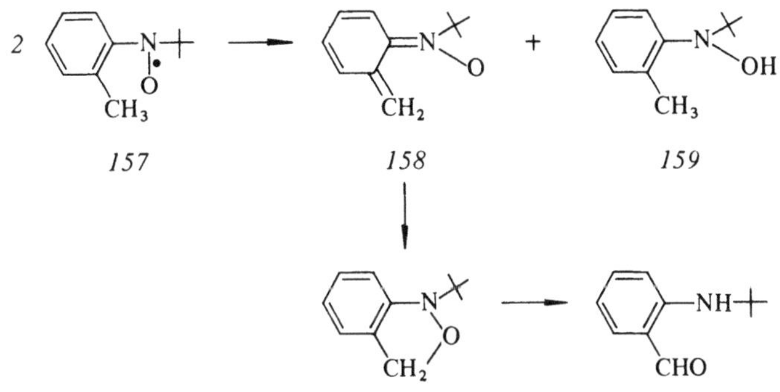

157 *158* *159* *160* *161*

Tert-butyl-p-vinylphenylaminyloxides as well as its ortho-analogs are extremely unstable, whereas the m-analogs are more stable, as expected. The p-vinyl-substituted radical spontaneously forms polymers of high molecular weights. In this case attack at the vinyl group by oxygen of a second radical molecule is assumed as the first step of the reaction[150].

c) Further Reactions at Aminyloxide Group

A further possibility for reactions of aminyloxides is addition of a Lewis-acid to the oxygen atom giving *de facto* an aminylradical *162*. ESR studies of complexes of aminyloxides with typical Lewis-acids BX_3[153], AlX_3[153, 154], GaX_3[154b], InX_3[155] (X = halogen) indicate spin delocalization to the complexing agent, the metal atoms showing sometimes considerable coupling to the unpaired electron. Such effects are also shown in complexes of aminyloxides with transition metals[115, 156].

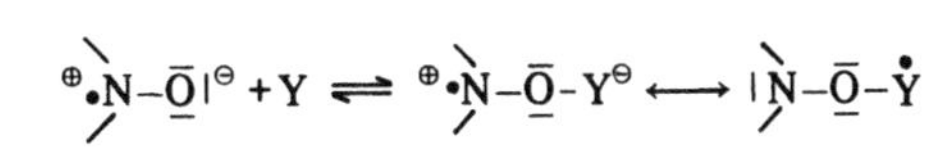

162

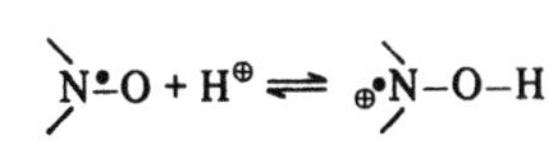

A special case of acid-base reaction is protonation of aminyloxides. Contrary to results of former studies[157] it was shown[158] that 2,2,6,6-tetramethylpiperidyloxide *3* in concentrated sulfuric acid or in trifluoroacetic acid forms the conjugate acid at room temperature, as do also di-tert-butlyaminyloxide *2b* and bicyclic aminyloxide *137*. It was found that dialkylaminyloxides behave as soft bases.

Formation of complexes of aminyloxides and organic ligands was studied, as for instance donor-acceptor-complexes with tetracyanoethylene[159]. Formation of inclusion compounds with aminyloxides was also observed[160].

Of special interest is the interaction of aminyloxides and solvents. A considerable influence of polarity of solvent to distribution of spin density was found. The interaction of aminyloxide group and protic solvents is defined by formation of hydrogen-bridge-bonds. For this hydrogen bridging the models A (σ-model) and B (π-model) are discussed by Buchachenko, the former being favored[161].

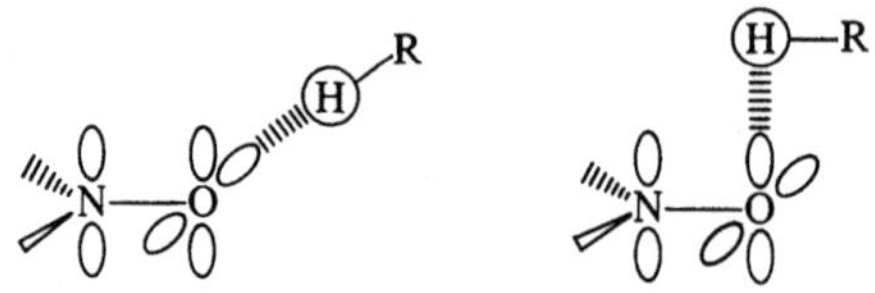

A further interesting goal is the ESR spectroscopic investigation of aminyloxides adsorbed on silica-aluminia showing the presence of two species, one according to protonation of aminyloxides, the other according to interaction with metal centers[162]. In disproportionation of aminyloxides in presence of acids[109a], reversible protonation is believed to be the first step of reaction[163, 157a]. Reaction products are immoniumoxide salt *163* and hydroxylammonium salt *164*. *163* is to be considered as a direct product of oxidation of aminyloxides. Suitable oxidizing reagents for its generation are chlorine[164], bromine[164, 165], antimonypentachloride[165b, 166] and triphenylcarbenium salts[167]. Secondary reactions with solvent may occur. For in-

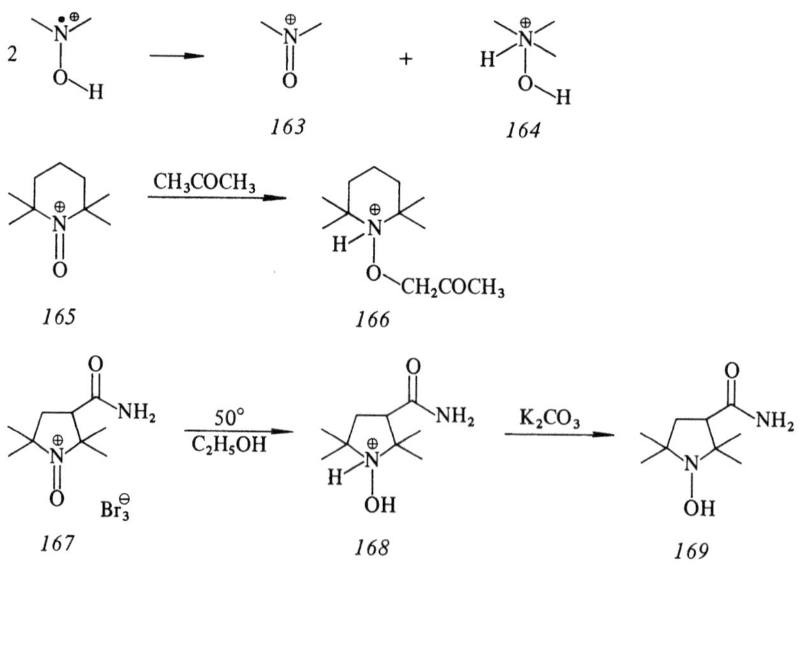

stance in acetone, hydroxylammonium salt *166* is formed from *165*[168]. Immonium-oxide salt *167* is reduced by warming in alcohol, giving *168* from which *169* may be obtained[165a].

Treatment of aminyloxides with a variety of reagents such as triethylphosphite[169], methyliodide or iodine[170], sodium sulfide in dimethylsulfoxide[171], tetramethyl-thiuransulfide[172] and sulfur dioxide[173] gives deoxygenation. But for triethylphosphite the mechanism of reaction has only been discussed. In the reaction of tert-butyl-phenylaminyloxide the first step is assumed to be the formation of a P–O-bond followed by scission of the N–O-bond, yielding triethylphosphate and an aminyl radical which finally abstracts hydrogen from solvent forming tert-butyl-phenylamine[169].

d) Reactions with β-Scission

Homolytic scission of a bond in the β-position of an alkylaminyloxide formally leads to formation of a nitrone and a radical. This reaction may be considered as reversal of spin trapping by nitrones. In these reactions the steric arrangement of substituents is very important. The most simple case is the oxidation of an aminyloxide *170* to

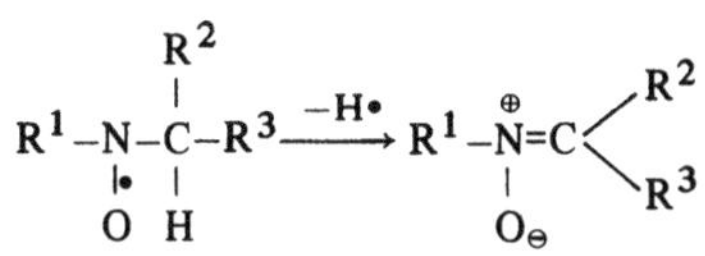

170 *171*

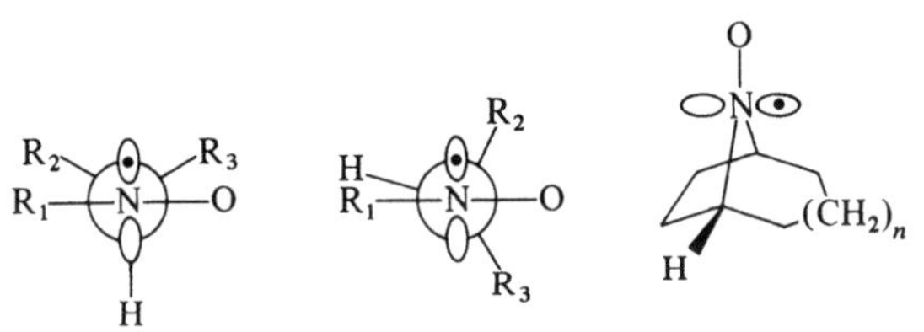

170A *170B* *172*

nitrone *171* breaking a β–C–H-bond. The removal of β-hydrogen is only possible from a conformation *170A* in which the C–H-bond lies in the same plane as the orbital containing the unpaired electron. This arrangement may be realized without steric hindrance in aminyloxides being substituted by methyl groups ($R^2 = R^3 = H$) or primary alkyl groups (R^2 or $R^3 = H$), explaining the easy disproportionation of radicals of this type (pp. 90/91). In aminyloxides containing secondary alkyl groups ($R^2, R^3 \neq H$) conformation *170A* becomes energetically more unfavorable with respect to other conformations according to enhanced steric interactions with the oxygen atom and substituent R^1. Thus such aminyloxides prefer conformation *170B* as was shown by the ESR coupling constants of the β-proton[174]. Hence these radicals are more stable and could even be isolated in some exceptional cases[175]. The fact that in bicyclic aminyloxides of type *172* a conformation like *170A* is made impossible by the rigid fixation of the aminyloxide group is an even more concrete explanation for the stability of these radicals than the Bredt rule.

On the other hand oxidation above the nitrone stage *171* is possible in aminyloxides like *173*. In these cases the intermediate nitrones *174* being potential tautomers of corresponding hydroxylamines (p. 69) are further dehydrogenated giving aminyloxides *175*. Thus aminyloxides *176* and *178* containing amino groups in the

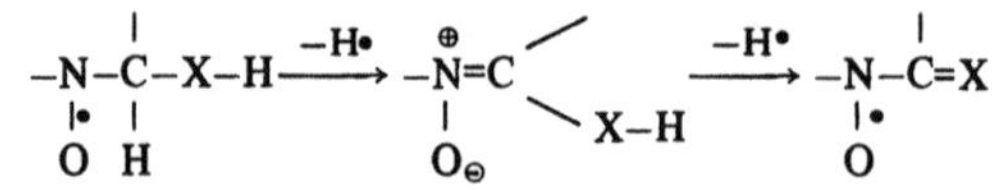

173 *174* *175*

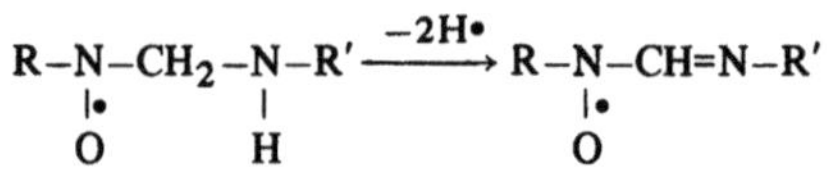

176 *177*

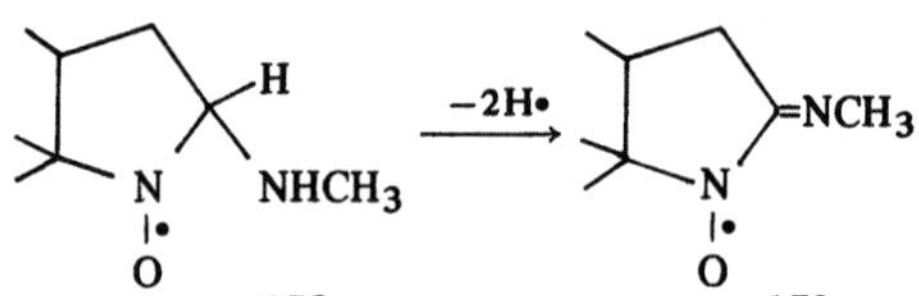

178 *179*

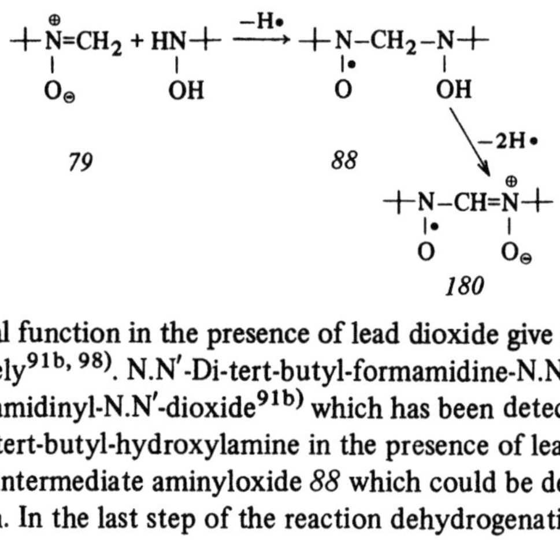

79 *88* *180*

β-position to the radical function in the presence of lead dioxide give aminyloxides *177* and *179* respectively[91b, 98]. N.N′-Di-tert-butyl-formamidine-N.N′-dioxide (*180*) being the first acyclic amidinyl-N.N′-dioxide[91b] which has been detected, is formed by reaction of *79* and tert-butyl-hydroxylamine in the presence of lead dioxide. This reaction goes through intermediate aminyloxide *88* which could be detected only with special precaution. In the last step of the reaction dehydrogenation of *88* leads to *180*[98].

Further oxidation of primary radicals also occurs in reaction of nitrile oxides *6* and N-substituted hydroxylamines *5* giving primary radicals *8* as was already mentioned[23] (pp. 69/70). From primary radical *181*, formation of *183* arises presumably *via* the intermediate nitrone *182* which suffers ring closure and further oxidation. The secondary radicals derived from *184* are *186* or *187* as determined by ESR spectroscopy. In this oxidation process no β-scission is possible but the δ-position possesses a hydrogen atom which may be abstracted. In fact formation of intermediate *185*[176] seems reasonable. Dependent on the size of the substituent R, ring closure of *185* and further oxidation leads to *186* or *187*. The possibility may not be excluded that in oxidation of *181* the first step is also dehydrogenation at the oxime group, nitrone *182* being formed subsequently by intramolecular hydrogen transfer. In all these reactions nitrile oxide *6* may operate as oxidizing agent so that no additional oxidizing agent is needed[23].

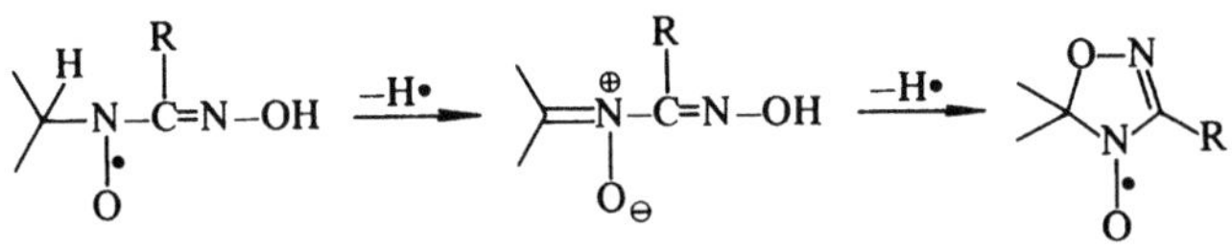

181 *182* *183*

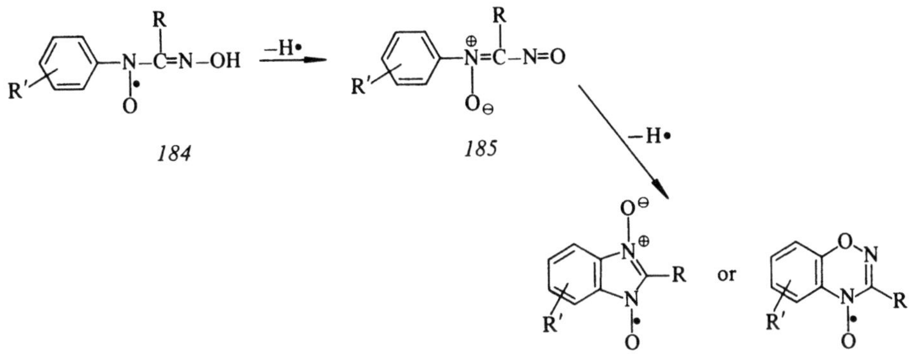

184 *185* *186* *187*

Besides scission of the β-carbon-hydrogen bond, scission of the β-carbon-halogen bond is also possible. In photolyses of α-chloronitroso compound *188* nitrone *190* is obtained[177]. Its formation is discussed *via* an intermediate aminyloxide *189* which, however, has not been detected by ESR spectroscopy. For the formation of chloro-

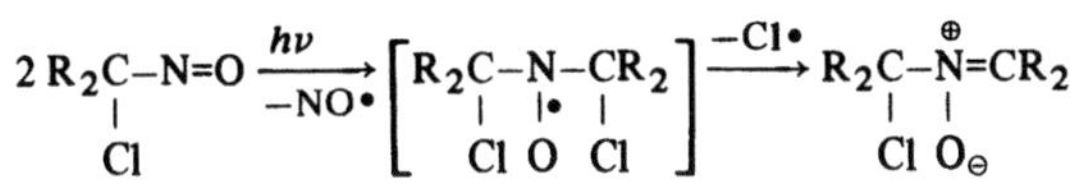

188 *189* *190*

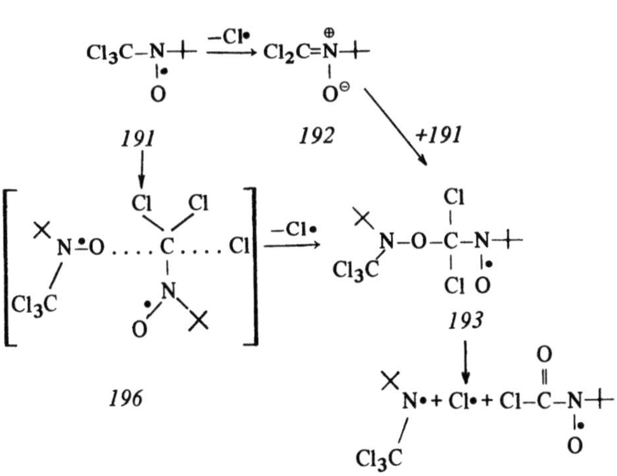

191 *192* *193* *196* *194* *195* *125a*

carbonylaminyloxides *125a* resulting from spin trapping of trichloromethyl radicals by 2-methyl-2-nitrosopropane[60, 117a, b] a reaction step including a β-scission of halogen from primarily formed trichloromethylaminyloxide *191* is discussed. The nitrone *192* thus formed is assumed to trap a second radical *191*, giving *193*. Intermediate *193* may decompose to chlorocarbonylaminyloxide *125a* which is detected by ESR spectroscopy while the radicals *194* and *195* are destroyed by secondary reactions. The last steps of this reaction resemble the decompositions of p-halogen-phenylaminyloxides (p. 92). Besides this elimination-addition mechanism for reaction sequence *191* → *192* → *193*, a S_N2-mechanism *via* transition state *196* is also discussed[151].

The conversion of radical *197* by certain nucleophiles is a very complex reaction presumably including a β-scission of a carbon-carbon bond. Aminyloxide *197* may add nucleophiles such as HCN, secondary aliphatic amines and ammonia at the carbonyl group forming aminyloxides *198*, which are detected by ESR spectroscopy, the addition of ammonia being reversible[179]. In reaction with hydroxide ions *198* could not be detected but the secondary radical *201*[178] is detected. On the other hand *201* may react with further nucleophiles giving radicals *202* by ring opening. In reaction of primary amines with aminyloxide *197*, only carbamoylaminyloxides analogs to *202*[c] could be detected, but it is reasonable to assume that the reaction also occurs *via* the corresponding intermediate *201*[c] which in this case is converted to *202*[c] more quickly. In sharp contrast to the course of these reactions are the reactions of nitrone *203*, the corresponding reduced compound. Treatment of *203* with the same nucleophiles leads in high yield to nitrones *205* by ring opening.

The aminyloxide *206* (X = OH), being formed by oxidation of *205* (X = OH) does not suffer ring closure to *201*[179]. Thus for reaction sequence *198* → *201*, an intermediate *206* resulting from heterolytic ring opening of *198* may be excluded. A reasonable course of reaction is, therefore, formation of an oxiran intermediate *199* which is converted subsequently to *200* by a β-scission of a C–C-bond. *200* is then oxidized giving radical *201*. This reaction also occurs in the absence of any additional oxidizing reagent, indicating that the oxidation *200* → *201* is caused by some portion of the aminyloxides in solution.

Conversion *198* → *201* is an obvious example for reactions occurring with involvement of aminyloxide function (β-scission, dehydrogenation) the radical function being regenerated at least. On the other hand, nucleophilic addition *197* → *198* and ring opening *201* → *202* occurs clearly without involvement of radical function.

A further example of reactions including a β-scission of the carbon-carbon bond is the photolysis of amidinyl-N.N′-dioxide *207* in water, the two radicals *212* and *215* being isolated by tlc[180]. This reaction may be considered to occur in the following steps: Intramolecular hydrogen abstraction is induced by light, giving radical *208* which suffers an intramolecular spin trapping forming aminyloxide *209*. β-Scission of the C–C-bond leads to the intermediate radical *210* which may either by reaction with oxygen form *212* *via* *211* or may add the starting radical *207* yielding intermediate *214* which fragmentates to *215* and *216*. The course of reaction through intermediate *210* is confirmed by photolysis of *213* yielding the same mixture of product radicals *212* and *215*.

---

[c] C=NR instead of C=O.

197 198

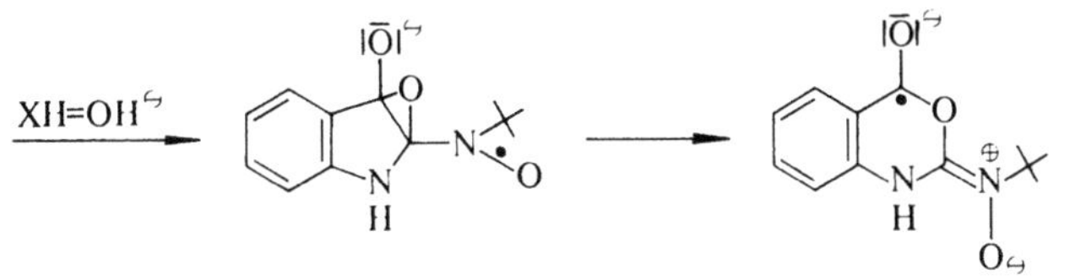

199 200

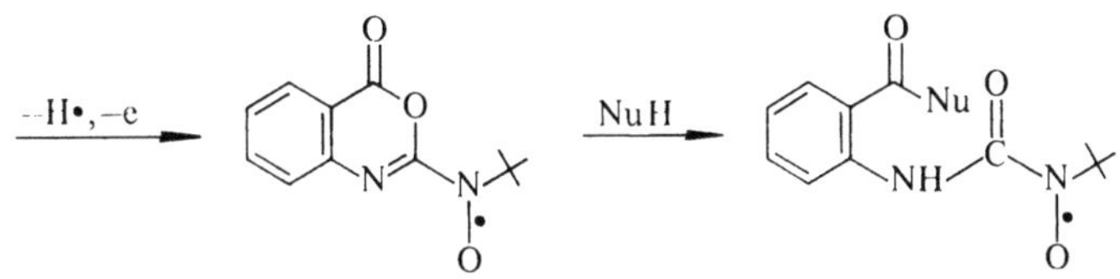

201 202

203 204

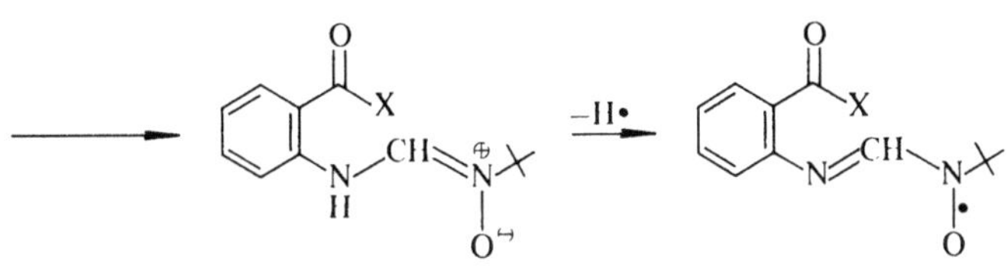

205 206

β-Scission of a carbon-oxygen bond in aminyloxides is supposed to occur in the conversion of formamidinyloxide *217* in the presence of carboxylic acids giving radicals *220* and *221*[181]. The reaction is assumed to start with the addition of carboxylic acid, as, for instance, acetic acid to C=N double bone. From intermediate *218* aminyloxide *220* is formed by homolytic rupture of the C—O-bond and a simultaneous attack of the developing acetoxy radical at the ortho-position of the phenyl ring giving intermediate *219* which stabilizes by loss of acetyl group, ring closure and further dehydrogenation. Surprisingly, conversion of radical *217* to *220* is strongly enhanced by freezing out the solvent. At room temperature formation of carbamoyl-aminyloxide *221* could be observed besides *220*. *221* is also thought to form through intermediate *218*.

It was confirmed by treatment of *217* with ^{18}O-labeled acetic acid and subsequent study of the reduced reaction mixture by mass spectrometry that oxygen from acetic acid was introduced into the benzoxazolyl ring of *220* and also into the carbonyl group of *221*[23a].

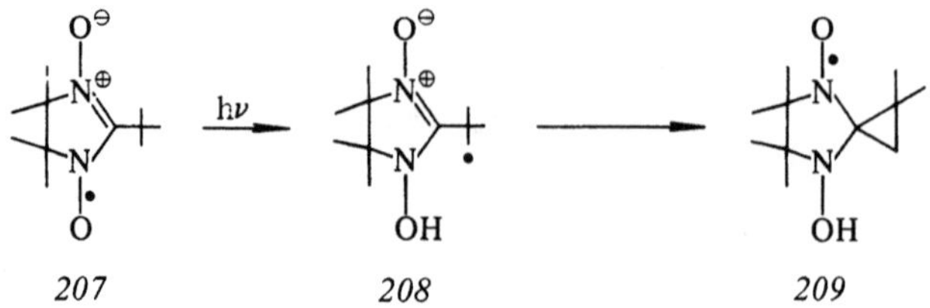
hν
207
208
209

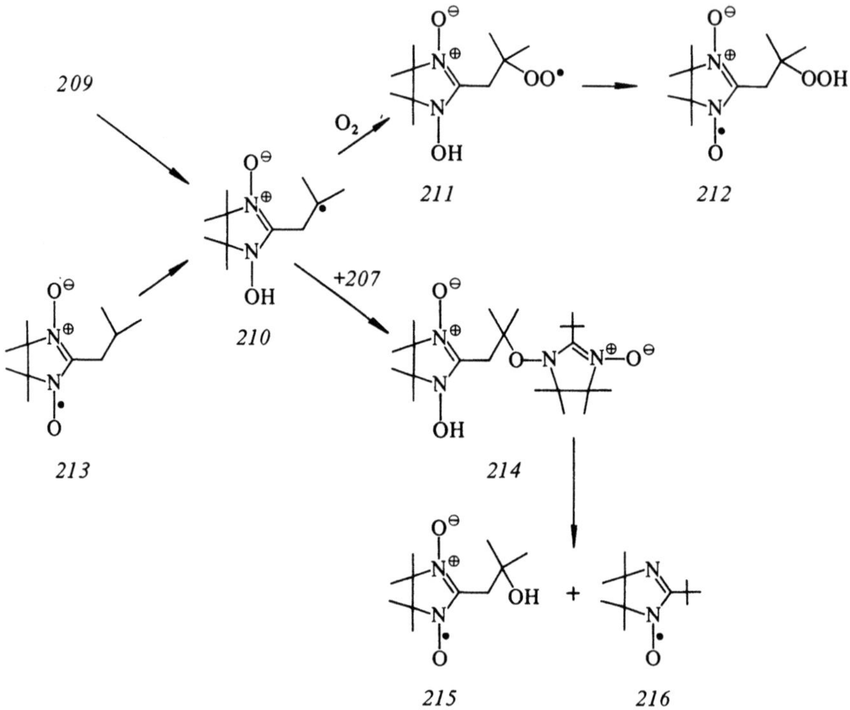
209
O₂
+207
210
211
212
213
214
215
216

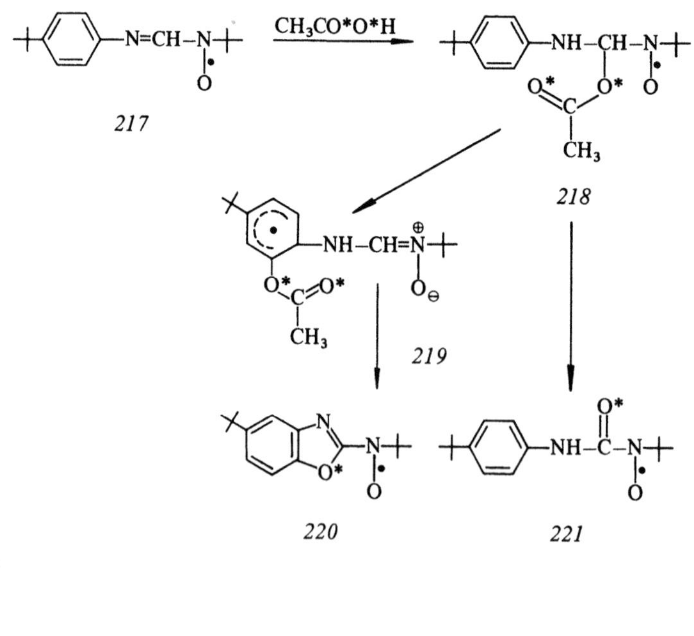
CH₃CO*O*H
217
218
219
220
221

e) Reactions with α-Scission

α-Scission of aminyloxides is the homolytic rupture of a bond of the nitrogen atom, formally leading to nitroso compound and a radical. This process is the reversal of spin trapping by nitroso compounds.

$$R{-}\underset{\substack{|\bullet \\ O}}{N}{-}R' \longrightarrow R{-}N{=}O + R'^{\bullet}$$

The α-scission of a N–C-bond occurs easily if a relatively stable carbon radical is formed. Thus trityl-aminyloxides *2* $R' = C(C_6H_5)_3$ decompose giving triphenylmethylradicals and nitroso compound[182]. The reaction takes place easily if the group R is bulky, as for instance tert-butyl.

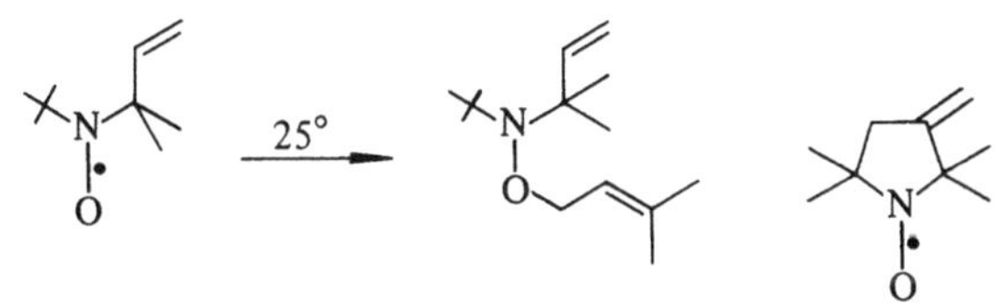

222 *223* *224*

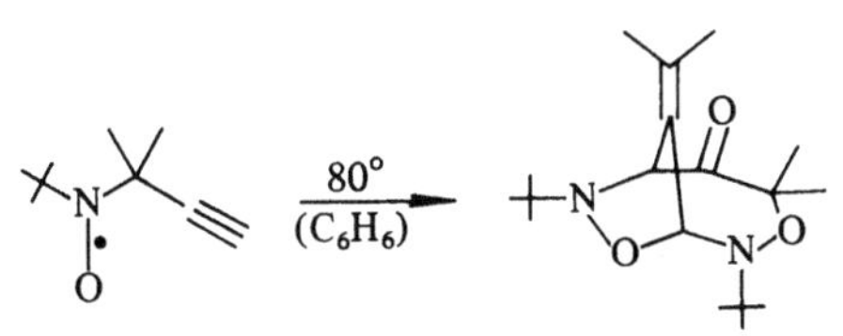

225 *226*

Decomposition products of the surprisingly unstable aminyloxide *222* are 2-methyl-2-nitrosopropane and hydroxylamine *223*[183]. The extraordinarily easy decomposition of *222* is caused by the formation of the energetically favorable dimethyl allyl radical by α-scission, which then adds further starting radical *222* giving *223*. On the other hand the cyclic aminyloxide *224* is very stable.

Propin-2yl-aminyloxide *225* decomposes only when refluxed in benzene solution giving bicyclic compound *226*, the structure of which was confirmed by X-ray analysis[184]. Apparently the first step of this reaction is again the α-scission of a N–C-bond giving 1.1-dimethyl-2-propinyl radical and 2-methyl-2-nitrosopropane, compound *226* being formed in an as yet unknown sequence of reactions.

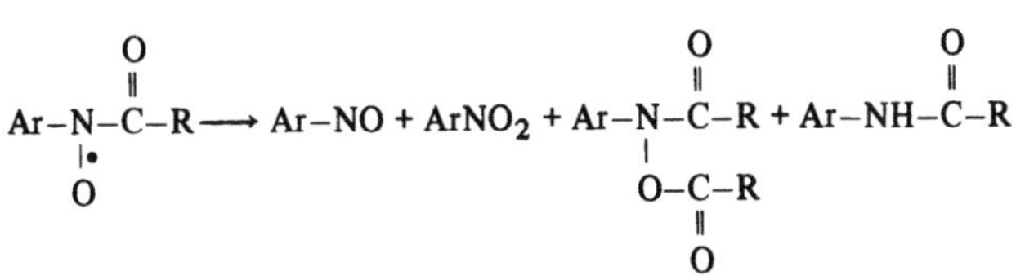

227 *228* *229* *230* *231*

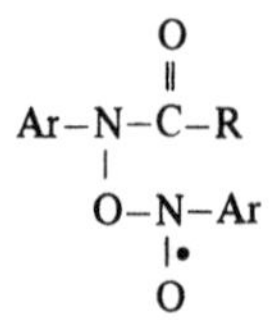

232

Studying the decomposition of acylaminyloxides *227* A. R. Forrester and co-workers[185)] isolated nitroso compounds *228*, nitro compounds *229*, N.O-diacyl-hydroxylamines *230* and amides *231*. Probably reaction is initiated by an α-scission giving *228* and an acyl radical. The latter may add further starting compound *227* yielding *230*. The formation of *229* and *231* may be explained through an intermediate radical *232* giving *229* and an acylaminyl radical by β-scission, which by hydrogen abstraction from solvent yields *231*. The decomposition of other carbonyl-aminyloxides has been studied also by Perkins[186)]. Janzen demonstrated further that the stability of N-aryl-benzoylaminyloxides (*227* R = Ar) is strongly affected by substituents at both arylrings, electron-withdrawing groups being destabilizing[19)].

The N–C-bond dissociation energy of di-tert-butyl-aminyloxide *2b* has been estimated to be 29 kcal/Mol[53)] according to the high thermal stability of *2b* which is decomposed by α-scission only at temperatures above 125 °C. In addition to 2-methyl-2-nitrosopropane, N.N.O-tri-tert-butylhydroxylamine was formed, the tert-butyl radical being trapped by unreacted *2b*.

Decomposition of 2.2.5.5-tetramethyl-3-carbamoyl-pyrrol-3-inyl-N-oxide by photolysis yielding 2-isopropylidene-4-methyl-3-pentenamide and nitric oxide may be considered as a two-fold α-scission of N–C-bond[187)].

Photolysis of amidinyl-N.N′-dioxide *207* in aprotic solvents is quite different from photolysis in water (pp. 98–100). It is a very complex reaction[188)] involving an α-scission of N–C-bond, leading to radicals *235* and *236*. Now the first step of the

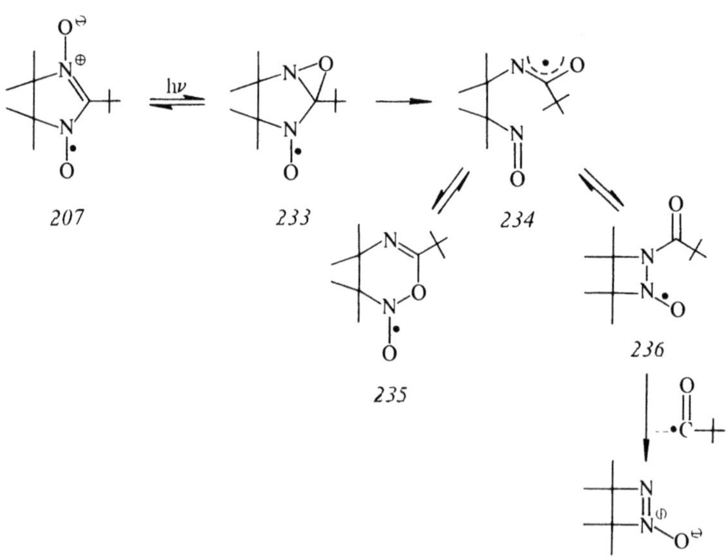

207 *233* *234* *235* *236* *237*

reaction is the formation of the oxaziridine ring which is typical for photochemistry of nitrones. Cleavage of the weak NO-bond of the three-membered ring in *233* initiates the further reactions. Primarily the α-scission at aminyloxide function leads to radical *234* which may either form radical *235* by intramolecular spin-trapping at oxygen or radical *236* by intramolecular spin-trapping at nitrogen. Radical *236* slowly converts to the thermodynamically more stable aminyloxide *235*. This conversion may be interpreted as α-scission of the N–N-bond regenerating *234* followed by intramolecular spin-trapping at oxygen. Heating the mixture of radicals *235* and *236* yields azoxy compound *237*. The loss of the pivaloyl group may be considered as scission of the N–C-bond in the β-position to the aminyloxide group.

The activation energy for α-scission of N–O-bond in alkoxyaminyloxides *69* has been determined as 10 kcal/Mol[78b]. This low value indicates the rather easy cleavage of radicals of type *69*. The reactions of the secondary radicals formed in this way have been discussed in Chapter III.4.

An interesting example of N.N-scission in the α-position to the aminyloxide-group is involved in the conversion of bicyclic hydrazyls *238* in the presence of oxygen[189]. Aminyloxide *239* resulting from oxidation of *238* slowly converts to *241*. For this process an α-scission of *239* is assumed, giving aminyl radical *240* which is oxidized to aminyloxide *241* in the following step.

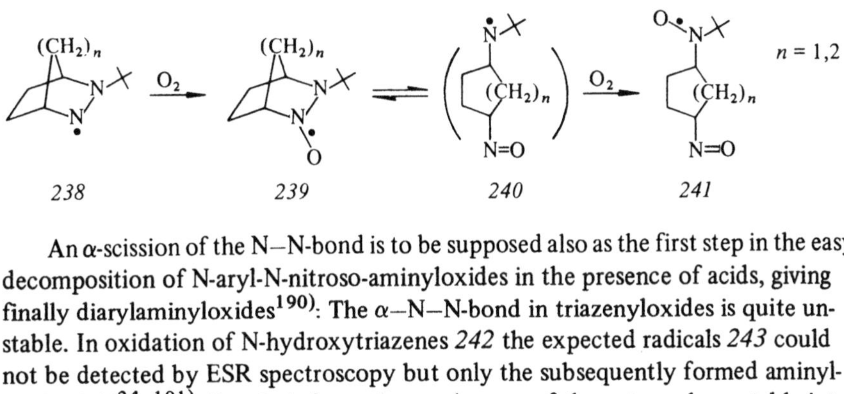

238 *239* *240* *241*

An α-scission of the N–N-bond is to be supposed also as the first step in the easy decomposition of N-aryl-N-nitroso-aminyloxides in the presence of acids, giving finally diarylaminyloxides[190]. The α–N–N-bond in triazenyloxides is quite unstable. In oxidation of N-hydroxytriazenes *242* the expected radicals *243* could not be detected by ESR spectroscopy but only the subsequently formed aminyloxides *244*[24, 191]. For their formation α-cleavage of the extremely unstable intermediate triazenyloxide *243* is assumed giving the nitroso compound and a diazonium radical which further decomposes by loss of nitrogen. The radical R· thus formed may be trapped by the nitroso compound giving aminyloxide *244*. It may also be possible that the splitting of the α-bond and loss of nitrogen occurs more or less simultaneously.

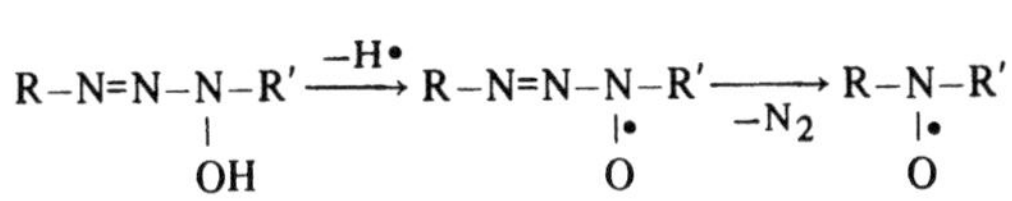

242 *243* *244*

On the other hand the N–N-bond of comparable aminyloxide *149* as well as the N–C-bond of comparable aminyloxides *177* is quite stable. Only in photolysis of *177* R = aryl R' = tert-butyl was formation of the corresponding radicals *244* observed to an extremely small extent[192], suggesting an analogous fragmentation with formation of H—C≡N under these conditions.

The first triazenyloxide being detected by ESR spectroscopy was benzotriazolyloxide *245*[193]. This radical also decomposes easily by loss of nitrogen forming secondary radicals by reaction with solvent or other suitable reagents[193, 194].

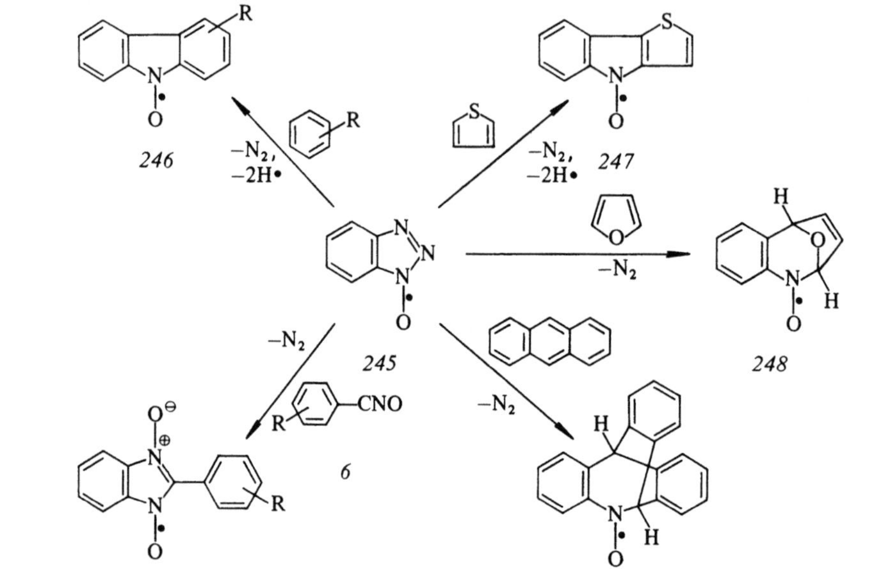

246 *247* *245* *248* *6* *249* *250*

For instance in benzene and some of its homologs, carbazolyl-N-oxides *246* are detected, in thiophene the heteroanalog radical *247* is formed. On the other hand in furan a 1.4-addition takes place giving product radical *248*. Analogously, reaction with anthracene leads to formation of *250*. Furthermore reaction of *245* with benzonitrile oxides *6* yields benzimidazolyl-N.N′-dioxides *249*.

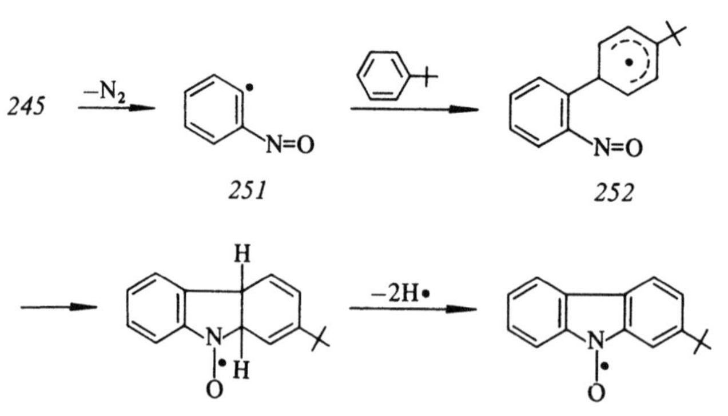

251 *252*

253 *254*

Reaction of *245* in tert-butylbenzene gives carbazolyl-N-oxide *254* containing the tert-butyl-group in the 2-position[193]. Thus it is assumed that loss of nitrogen is the first step of reaction giving nitrosophenyl radical *251*. *251* attacks the p-position of tert-butylbenzene with formation of *252*. Ring closure by intramolecular spin trapping followed by dehydrogenation yields *254 via 253*. The fact that in the absence of additional oxidizing agents reaction also occurs indicates that, in this case, the oxidation is carried out by radicals in solution.

Photolysis of 2-iodo-nitrosobenzene in benzene also yields radical *246* ($\dot{R}$ = H) confirming the intermediate appearance of *251*[194].

Formation of radicals *248* and *250* may be considered as (3 + 4)-cycloaddition involving the nitrosophenyl radical *251*.

V. Summary

The low energy content of the aminyloxide group is responsible for the easy formation of aminyloxides not only by a variety of simple reactions but also by many complex reactions. Besides various reactions of aminyloxides leading to the disappearance of radical function there are many reactions occurring at other functional groups in the molecule without the involvement of the aminyloxide function, hence retaining the radical character.

Only recently another type of reaction has been more extensively studied in which radical function is involved but finally retained. In these reactions the aminyloxide function disappears only temporarily during the course of reaction with subsequent regeneration. In contrast to reactions without involvement of the aminyloxide group these reactions should be named as reactions with participation and preservation of the aminyloxide function.

Acknowledgements. Our work, mentioned in this paper, was generously supported by the Deutsche Forschungsgemeinschaft and the Fonds der Chemischen Industrie.

We wish to thank Prof. Dr. A. Berndt for critical comments and helpful discussions.

I (H. G. A.) should like to thank my co-workers for their contributions to the work mentioned in this review.

VI. References

1) a) Piloty, O., Graf Schwerin, B.: Ber. Deut. Chem. Ges. *34*, 1870 (1901);
b) Piloty, O., Graf Schwerin, B.: Ber. Deut. Chem. Ges. *34*, 2354 (1901).
2) Wieland, H., Offenbächer, M.: Ber. Deut. Chem. Ges. *47*, 2111 (1914).
3) Rozantsev, E. G.: Free nitroxyl radicals. New York–London: Plenum Press 1970.
4) a) Rozantsev, E. G., Sholle, V. D.: Synthesis *1971*, 190;
b) Rozantsev, E. G., Sholle, V. D.: Synthesis *1971*, 401.
5) Forrester, A. R., Hay, J. M., Thomson, R. H.: Organic chemistry of stable free radicals, p. 180. London and New York: Academic Press 1968; especially page 191.
6) Rozantsev, E. G.: Free nitroxyl radicals. New York–London: Plenum Press 1970 (especially pages 120–125).

7) Hayat, H., Silver, B. L.: J. Phys. Chem. *77*, 72 (1973).
8) Roberts, J. R., Ingold, K. U.: J. Am. Chem. Soc. *95*, 3228 (1973).
9) a) Cohen, A. H., Hoffman, B. M.: J. Phys. Chem. *78*, 1313 (1974);
b) Cohen, A. H., Hoffman, B. M.: J. Am. Chem. Soc. *95*, 2061 (1973).
10) Faber, R. J.: J. Chem. Phys. *46*, 1613 (1967).
11) Baird, J. C.: J. Chem. Phys. *37*, 1879 (1962).
12) Aurich, H. G., Hahn, K., Stork, K., Weiss, W.: unpublished results.
13) Abronin, I. A., Sysoeva, N. A., Zhidomirov, G. M., Buchachenko, A. L., Chuvylkin, N. D.: Zh. Strukt. Khim. *12*, 985 (1971); J. Struct. Chem. USSR (Engl. Transl.), *12*, 907 (1971).
14) Andersen, B., Andersen, P.: Acta Chem. Scand. *20*, 2728 (1966).
15) Berliner, L. J.: Acta Crystallogr. *B 26*, 1198 (1970).
16) Reichardt, Ch., Dimroth, K.: Fortschr. chem. forsch. *11*, 1 (1968/1969); (especially page 50).
17) Aurich, H. G., Trösken, J.: Liebigs Ann. Chem. *745*, 159 (1971).
18) Aurich, H. G., Dersch, W., Forster, H.: Chem. Ber. *106*, 2854 (1973).
19) Janzen, E. G.: Acc. Chem. Res. *2*, 279 (1969).
20) Blackley, W. D., Reinhard, R. R.: J. Am. Chem. Soc. *87*, 802 (1965).
21) *loc. cit.*[4a] page 190 and *cit.*[5] page 180 f.
22) Mahoney, L. R., Mendenhall, G. D., Ingold, K. U.: J. Am. Chem. Soc. *95*, 8611 (1973).
23) a) Stork, K.: Diss. Univ. Marburg 1974;
b) Aurich, H. G., Stork, K.: Chem. Ber. *108*, 2764 (1975).
24) Aurich, H. G., Baer, F.: Tetrahedron Letters *1965*, 3879.
25) a) Hahn, K.: Dipl.-Arbeit, Univ. Marburg 1974;
b) Aurich, H. G., Hahn, K., Stork, K.: Angew. Chem. *87*, 590 (1975); Angew. Chem. Internat. Edit. *14*, 551 (1975).
26) Aurich, H. G., Dersch, W., Blinne, G.: Liebigs Ann. Chem. *762*, 154 (1972).
27) Cowley, D. J., Waters, W. A.: J. Chem. Soc. (B), *1970*, 96.
28) Cowley, D. J., Millen, M. H., Waters, W. A.: J. Chem. Soc. (B), *1971*, 2393.
29) Millen, M. H., Waters, W. A.: J. Chem. Soc. (B), *1971*, 2398.
30) West, R., Boudjouk, P.: J. Am. Chem. Soc. *93*, 5901 (1971).
31) *loc. cit.*[4a] and *cit.*[5]: page 185.
32) Toda, T., Mori, E., Horiuchu, H., Murayama, K.: Bull. Chem. Soc. Jap. *45*, 1802 (1972).
33) a) Neugebauer, F. A., Fischer, P. H. H.: Chem. Ber. *98*, 844 (1965);
b) Fischer, P. H. H., Neugebauer, F. A.: Z. Naturforsch. *19A*, 1514 (1964).
34) Nelsen, S. F., Landis, II, R. T.: J. Am. Chem. Soc. *95*, 8707 (1973).
35) Danen, W. C., West, C. T., Kensler, T. T.: J. Am. Chem. Soc. *95*, 5716 (1973).
36) *loc. cit.*[3], page 67.
37) a) Boozev, Ch. E., Hammond, G. S.: J. Am. Chem. Soc. *76*, 3861 (1954);
b) Thomas, J. R., Tolman, C. A.: J. Am. Chem. Soc. *84*, 2930 (1962).
38) Janzen, E. G.: Acc. Chem. Res. *4*, 31 (1971).
39) Forrester, A. R., Hepburn, S. P.: J. Chem. Soc. (C), *1971*, 701.
40) Holman, R. J., Perkins, M. J.: J. Chem. Soc. (C), *1971*, 2324.
41) Lagercrantz, C., Torsell, K.: Acta Chem. Scand. *22*, 1935 (1965).
42) Hartgerink, J. W., Engberts, J. B. F. N., Wajer, Th. A. J., de Boer, Th. J.: Rec. Trav. Chim. Pays-Bas *88*, 481 (1969).
43) Stammer, R., Engberts, J. B. F. N., de Boer, Th. J.: Rec. Trav. Chim. Pays-Bas *89*, 169 (1970).
44) Terabe, S., Kuruma, K., Konaka, R.: JCS Perkin II, *1973*, 1252.
45) a) Terabe, S., Konaka, R.: JCS Perkin II, *1973*, 369;
b) Terabe, S., Konaka, R.: Bull. Chem. Soc. Jap. *46*, 825 (1973).
46) a) Ahrens, W., Wieser, K., Berndt, A.: unpublished results;
b) Ahrens, W., Wieser, K., Berndt, A.: Tetrahedron Letters *1973*, 3141.
47) a) Cheng, I.-D., Shine, H. J.: J. Org. Chem. *39*, 336 (1974);
b) Edwards, O. E., Paskovich, D. H., Reddoch, A. H.: Can. J. Chem. *51*, 978 (1973);
c) Terabe, S., Konaka, R.: JCS Perkin II, *1972*, 2163;
d) Terabe, S., Konaka, R.: J. Am. Chem. Soc. *91*, 5655 (1969);

e) Lagercrantz, C., Forshult, S.: Acta Chem. Scand. *23*, 708 (1969);
f) Leaver, J. H., Ramsay, G. C., Suzuki, E.: Austr. J. Chem. *22*, 1891 (1969);
g) Leaver, J. H., Ramsay, G. C.: Austr. J. Chem. *22*, 1899 (1969).

48) a) Mackor, A., Wajer, Th. A. J. W., de Boer, Th. J.: Tetrahedron Letters *1967*, 385;
b) Wargon, J. A., Williams, F.: J. Am. Chem. Soc. *94*, 7917 (1972);
c) Forshult, S., Lagercrantz, C.: Acta Chem. Scand. *23*, 522 (1969);
d) Perkins, M. J., Ward, P., Horsfield, A.: J. Chem. Soc. (B), *1970*, 395;
e) Abakumov, G. A., Razuvaev, G. A.: Dokl. Akad. Nauk SSSR *182*, 95 (1968); C. A. *70*, 2968 (1969).

49) a) Karlsson, H., Lagercrantz, C.: Acta Chem. Scand. *24*, 3411 (1970);
b) Adevik, G., Lagercrantz, C.: Acta Chem. Scand. *24*, 2253 (1970);
c) Boekestein, G., Buck, H. M.: Rec. Trav. Chim. Pays-Bas *92*, 1095 (1973).

50) Mulvey, D., Waters, W. A.: JCS Perkin II, *1974*, 666.

51) Mackor, A., Wajer, Th. A. J. W., de Boer, Th. J., van Voorst, J. D. W.: Tetrahedron Letters *1966*, 2115.

52) Wajer, Th. A. J. W., Mackor, A., de Boer, Th. J., van Voorst, J. D. W.: Tetrahedron *23*, 4021 (1967).

53) Carmichael, P. J., Gowenlock, B. G., Johnson, C. A. F.: JCS Perkin II, *1973*, 1853.

54) Maassen, J. A., Hittenhausen, H., de Boer, Th. J.: Tetrahedron Letters *1971*, 3213.

55) Rassat, A., Rey, P.: Chem. Commun. *1971*, 1161.

56) a) Malatesta, V., Ingold, K. U.: Tetrahedron Letters *1973*, 3307;
b) Maassen, J. A., de Boer, Th. J.: Rec. Trav. Chim. Pays-Bas *92*, 185 (1973);
c) Aurich, H. G., Dersch, W.: Z. Naturforsch. *28B*, 525 (1973).

57) Chalfont, G. R., Perkins, M. J., Horsfield, A.: J. Am. Chem. Soc. *90*, 7141 (1968).

58) Jonkman, L., Muller, H., Kommandeur, J.: J. Am. Chem. Soc. *93*, 5833 (1971).

59) Motherwell, W. B., Roberts, J. S.: JCS Chem. Commun. *1972*, 328.

60) Camaggi, C. M., Holman, R. J., Perkins, M. J.: JCS Perkin II, *1972*, 501.

61) Cadogan, J. I. G.: Acc. Chem. Res. *4*, 186 (1971).

62) Chalfont, G. R., Perkins, M. J.: J. Am. Chem. Soc. *89*, 3054 (1967).

63) Wajer, Th. A. J. W., Mackor, A., de Boer, Th. J.: Tetrahedron *25*, 175 (1969).

64) Russell, G. A., Geels, E. J., Smentowski, F. J., Chang, K.-Y., Reynolds, J., Kaupp, G.: J. Am. Chem. Soc. *89*, 3821 (1967).

65) Mulvey, D., Waters, W. A.: JCS Perkin II, *1974*, 772.

66) *loc. cit.*[5], page 181.

67) Ashby, E. C., Laemmle, J., Neumann, H. M.: Acc. Chem. Res. *7*, 272 (1974).

68) a) Regenstein, H.: Diss. Univ. Marburg 1974;
b) Berndt, A., Regenstein, H.: unpublished results.

69) Aurich, H. G., Baer, F.: Angew. Chem. *79*, 1070 (1967); Angew. Chem. Internat. Edit. *6*, 1080 (1967).

70) a) Sullivan, A. B.: J. Org. Chem. *31*, 2811 (1966);
b) Knight, G. T.: Chem. Commun. *1970*, 1016.

71) Banks, R. E., Haszeldine, R. N., Miller, P. J.: Tetrahedron Letters *1970*, 4417.

72) Layer, R. W.: Tetrahedron Letters *1970*, 4413.

73) a) Wassmuth, G.: Diss. Univ. Marburg 1974;
b) Aurich, H. G., Wassmuth, G.: unpublished results.

74) a) Lewis, J. W., Myers, P. L., Ormerod, J. A.: JCS Perkin I, *1972*, 2521;
b) Lewis, J. W., Myers, P. L. Ormerod, J. A.: JCS Perkin I, *1973*, 1129.

75) Wajer, Th. A. J. W., Geluk, H. W., Engberts, J. B. F. N., de Boer, Th. J.: Rec. Trav. Chim. Pays-Bas *89*, 696 (1970).

76) Lakatos, B., Turcsanyi, B., Tüdos, F.: Acta Chim. Acad. Sci. Hung, *70*, 225 (1971); C. A. *76*, 125960 (1972).

77) Banks, R. E., Edge, D. J., Freear, J., Haszeldine, R. N.: JCS Perkin I, *1974*, 721.

78) a) Janzen, E. G., Gerlock, J. L.: J. Am. Chem. Soc. *91*, 3108 (1969);
b) Cowley, D. J., Sutcliffe, L. H.: J. Chem. Soc. (B), *1970*, 569,

79) a) McMillan, M., Norman, R. O. C.: J. Chem. Soc. (B), *1968*, 590;

b) Anderson, N. H., McMillan, M., Norman, R. O. C.: ibid. *1970*, 1075.
80) Chachaty, C., Forchioni, A.: Tetrahedron Letters *1968*, 307.
81) Chachaty, C., Forchioni, A.: Tetrahedron Letters *1968*, 1079.
82) a) Ahrens, W., Berndt, A.: Tetrahedron Letters *1973*, 4281;
b) Ahrens, W., Berndt, A.: unpublished results;
c) Camaggi, C. M., Lunazzi, L., Placucci, G.: J. Org. Chem. *39*, 2425 (1974);
d) Camaggi, C. M., Leardini, R., Placucci, G.: JCS Perkin II, *1974*, 1195;
e) Camaggi, C. M., Lunazzi, L., Pedulli, G. F., Placucci, G., Tiecco, M.: JCS Perkin II, *1974*, 1226.
83) Chalfont, G. R., Hey, D. H., Liang, K. S. Y., Perkins, M. J.: J. Chem. Soc. (B), *1971*, 233.
84) Beckwith, A. L. J., Lawton, M. D.: JCS Perkin II, *1973*, 2134.
85) a) Hoffmann, A. K., Feldmann, A. M., Gelblum, E., Hodgson, W. G.: J. Am. Chem. Soc. *86*, 639 (1964);
b) Hoffmann, A. K., Hodgson, W. G., Maricle, D. L., Jura, W. H.: J. Am. Chem. Soc. *86*, 631 (1964).
86) Hoffmann, A. K., Feldmann, A. M., Gelblum, E.: J. Am. Chem. Soc. *86*, 646 (1964).
87) *loc. cit.*[5], page 182.
88) Perkins, M. J. in: Essays on free-radical chemistry (ed. R. O. C. Norman): Chem. Soc. Special Publ. *24*, 97 (1970).
89) Janzen, E. G., Blackburn, B. J.: J. Am. Chem. Soc. *91*, 4481 (1969).
90) a) Bluhm, A. L., Weinstein, J.: J. Org. Chem. *37*, 1748 (1972);
b) Janzen, E. G., Evans, C. A.: J. Am. Chem. Soc. *95*, 8205 (1973);
c) Janzen, E. G., Evans, C. A., Nishi, Y.: J. Am. Chem. Soc. *94*, 8236 (1972).
91) a) Chalfont, G. R., Perkin, M. J., Horsfield, A.: J. Chem. Soc. (B), *1970*, 401;
b) Aurich, H. G., Höhlein, P.: Tetrahedron Letters *1974*, 279.
92) a) Iwamura, M., Inamoto, N.: Bull. Chem. Soc. Jap. *40*, 703 (1967);
b) Iwamura, M., Inamoto, N.: Bull. Chem. Soc. Jap. *43*, 860 (1970);
c) Janzen, E. G., Liu, J. I.-P.: J. Magn. Res. *9*, 510 (1973);
d) Janzen, E. G., Evans, C. A., Liu, J. I.-P.: J. Magn. Res. *9*, 513 (1973).
93) Tyurikov, V. A., Shapiro, B. I., Neimann, L. A.: Zh. Org. Khim. *7*, 2372 (1971); J. Org. Chem. USSR (Engl. Transl.) *7*, 2463 (1971).
94) Janzen, E. G., Lopp, J. G.: J. Magn. Res. *7*, 107 (1972).
95) Bard, A. J., Gilbert, J. C., Goodin, R. D.: J. Am. Chem. Soc. *96*, 620 (1974).
96) Pacifici, J. G., Browning, Jr., H. L.: J. Am. Chem. Soc. *92*, 5231 (1970).
97) Camaggi, C. M., Perkins, M. J.: JCS Perkin II, *1972*, 507.
98) a) Höhlein, P.: Diss. Univ. Marburg 1974;
b) Aurich, H. G., Höhlein, P.: unpublished results.
99) Bluhm, A. L., Weinstein, J.: J. Am. Chem. Soc. *92*, 1444 (1970).
100) Sommermeyer, K., Seiffert, W., Wilker, W.: Tetrahedron Letters *1974*, 1821.
101) Eckersley, A., Rogers, N. A. J.: Tetrahedron Letters *1974*, 1661.
102) a) Caragheorgheopol, A., Căldăraru, H., Constantinescu, T., Em Sahini, V.: J. Am. Chem. Soc. *93*, 6766 (1971);
b) Lown, J. B.: J. Chem. Soc. (B), *1966*, 441;
c) Lown, J. B.: J. Chem. Soc. (B), *1966*, 644.
103) Wajer, Th. A. J. W., Mackor, A., de Boer, Th. J., van Voorst, J. D. W.: Tetrahedron *23*, 4021 (1967).
104) a) Smith, P., Fox, W. M.: Can. J. Chem. *47*, 2227 (1969);
b) Adams, J. Q.: J. Am. Chem. Soc. *89*, 6022 (1967).
105) Khadzhiev, D. S., Todorova-Khadzhieva, E. S.: Dokl. Bolg. Akad. Nauk *25*, 507 (1972); C. A. *77*, 114 226 (1972).
106) review article: Griffith, O. H., Waggoner, A. S.: Acc. Chem. Res. *2*, 17 (1969).
107) Gramicidin, S., Ivanov, V. T., Miroshnikov, A. I., Snezhkova, L. G., Ovchinnikov, Yu. A., Kulikov, A. V., Likhtenshtein, G. I.: Khim. Prir. Soedin. *9*, 91 (1973); C. A. *79*, 19077 (1973).
108) Osiecki, J. H., Ullman, E. F.: J. Am. Chem. Soc. *90*, 1078 (1968).
109) a) Boocock, D. G. B., Darcy, R., Ullman, E. F.: J. Am. Chem. Soc. *90*, 5945 (1968);
b) Boocock, D. G. B., Ullman, E. F.: J. Am. Chem. Soc. *90*, 6873 (1968).
110) Ullman, E. F., Boocock, D. G. B.: Chem. Commun. *1969*, 1161.
111) Ullman, E. F., Call, L., Leute, R. K., Osiecki, J. H.: U. S. 3, 740, 412; C. A. *79*, 42502 (1973).

112) Darcy, R., Ullman, E. F.: J. Am. Chem. Soc. *91*, 1024 (1969).
113) Ullman, E. F., Osiecki, J. H., Boocock, D. G. B., Darcy, R.: J. Am. Chem. Soc. *94*, 7049 (1972).
114) Ullman, E. F., Call, L., Osiecki, J. H.: J. Org. Chem. *35*, 3623 (1970).
115) Wagner, B. E., Helbert, J. N., Bates, Jr., R. D., Poindexter, E. H.: JCS Chem. Commun. *1973*, 748.
116) Medzhidov, A. A., Kirichenko, L. N., Likhtenshtein, G. I.: Izv. Akad. Nauk SSSR, Ser. Khim. *1969*, 698; C. A. *71*, 49713 (1969).
117) a) Hartgerink, J. W., Engberts, J. B. F. N., de Boer, Th. J.: Tetrahedron Letters *1971*, 2709;
b) Holman, R. J., Perkins, M. J.: Chem. Commun. *1971*, 244;
c) Aurich, H. G., Stork, K.: Tetrahedron Letters *1972*, 555.
118) a) Weiss, W.: Diss. Univ. Marburg 1973;
b) Aurich, H. G., Weiss, W.: Tetrahedron in press.
119) Arndt, F., Eistert, B., Pertale, W.: Ber. Deut. Chem. Ges. *60*, 1364 (1927).
120) *loc. cit.*[5], especially page 221 ff.
121) Sholle, V. D., Golubev, V. A., Rozantsev, E. G.: Dokl. Akad. Nauk SSSR *200*, 137 (1971); Dokl. Chem. (Engl. Transl.) *200*, 761 (1971).
122) Robbins, W. K., Eastman, R. H.: J. Am. Chem. Soc. *92*, 6077 (1970).
123) Rykov, S. V., Sholle, V. D.: Izv. Akad. Nauk SSSR, Ser. Khim. *1971*, 2351; C. A. *76*, 45448 (1972).
124) *loc. cit.*[4b]; (especially page 406).
125) Keana, J. F. W., Dinerstein, R. J., Baitis, F.: J. Org. Chem. *36*, 209 (1971).
126) a) Murayama, K., Yoshioka, T.: Bull. Chem. Soc. Jap. *42*, 2307 (1969);
b) Yoshioka, T., Higashida, S., Morimura, S., Murayama, K.: Bull. Chem. Soc. Jap. *44*, 2207 (1971).
127) a) Schlude, H.: Tetrahedron *29*, 4007 (1973);
b) Kabs, K.: Diss. Univ. Marburg 1970.
128) Toda, T., Mori, E., Murayama, K.: Bull. Chem. Soc. Jap. *45*, 1904 (1972).
129) Banks, R. E., Haszeldine, R. N., Stevenson, M. J.: J. Chem. Soc. (C), *1966*, 901.
130) a) Banks, R. E., Justin, B., Haszeldine, R. N.: J. Chem. Soc. (C), *1971*, 2777;
b) Banks, R. E., Chondhury, D. R., Haszeldine, R. N.: JCS Perkin I, *1973*, 1092.
131) Banks, R. E., Chondhury, D. R., Haszeldine, R. N.: JCS Perkin I, *1973*, 80.
132) Banks, R. E., Haszeldine, R. N., Myerscough, T.: JCS Perkin I, *1972*, 2336.
133) Banks, R. E., Haszeldine, R. N., Myerscough, T.: J. Chem. Soc. (C) *1971*, 1951.
134) Banks, R. E., Haszeldine, R. N., Stephens, C. W.: Tetrahedron Letters *1972*, 3699.
135) Banks, R. E., Cheng, W. M., Haszeldine, R. N., Show, G.: J. Chem. Soc. (C), *1970*, 55.
136) Mendenhall, G. D., Ingold, K. U.: J. Am. Chem. Soc. *95*, 6390 (1973).
137) Capiomont, A., Chion, B., Lajzérowicz, J.: Acta Cristallogr. *B27*, 322 (1971).
138) Ramasseul, R., Rassat, A., Rio, G., Scholl, M.-J.: Bull. Soc. Chim. Fr. *1971*, 215.
139) Moser, W., Howie, R. A.: J. Chem. Soc. (A), *1968*, 3039.
140) Mendenhall, G. D., Ingold, K. U.: J. Am. Chem. Soc. *95*, 6395 (1973).
141) a) Dupeyre, R. M., Rassat, A.: J. Am. Chem. Soc. *88*, 3180 (1966);
b) loc. cit[4b], page 401.
142) a) Adamic, K., Bowman, D. F., Ingold, K. U.: J. Am. Chem. Soc. *92*, 1093 (1970);
b) Adamic, K., Bowman, D. F., Gillan, T., Ingold, K. U.: J. Am. Chem. Soc. *93*, 902 (1971).
143) Bowman, D. F., Brokenshire, J. L., Gillan, T., Ingold, K. U.: J. Am. Chem. Soc. *93*, 6551 (1971).
144) Bowman, D. F., Gillan, T., Ingold, K. U.: J. Am. Chem. Soc. *93*, 6555 (1971).
145) Martinie-Hombrouck, J., Rassat, A.: Tetrahedron *30*, 433 (1974).
146) a) Woodward, R. B., Wintner, C.: Tetrahedron Letters *1969*, 2693;
b) Lotz, I.: Dipl.-Arbeit, Univ. Marburg 1974;
c) Aurich, H. G., Lotz, I., Weiss, W.: unpublished results.
147) Wieland, H., Roth, K.: Ber. Deut. Chem. Ges. *53*, 210 (1920).
148) Calder, A., Forrester, A. R.: Chem. Commun. *1967*, 682.
149) Forrester, A. R., Hepburn, S. P.: J. Chem. Soc. (C), *1970*, 1277.
150) Forrester, A. R., Hepburn, S. P.: J. Chem. Soc. (C), *1971*, 3322.
151) Calder, A., Forrester, A. R., Hepburn, S. P.: JCS Perkin I, *1973*, 456.
152) Abramovitch, R. A., Challand, S. R.: J. Heterocycl. Chem. *10*, 683 (1973).
153) a) Eames, T. B., Hoffman, B. M.: J. Am. Chem. Soc. *93*, 3141 (1971);
b) Cazianis, C. T., Eaton, D. R.: Can. J. Chem. *52*, 2463 (1974).
154) a) Hoffman, B. M., Eames, T. B.: J. Am. Chem. Soc. *91*, 5168 (1969);

b) Abakumov, G. A., Tikhonov, V. D., Razuyald, G. A.: Dokl. Akad. Nauk SSSR *187*, 571 (1969); C. A. *71*, 85129 (1969);
c) Razuvaev, G. A., Tikhonov, V. D., Abakumov, G. A.: Izv. Akad. Nauk SSSR, Ser. Khim. *1970*, 1732; C. A. *74*, 59210 (1971).
155) Abakumov, G. A., Tikhonov, V. D.: Zh. Strukt. Khim. *13*, 150 (1972); J. Strukt. Chem. USSR (Engl. Transl.) *13*, 136 (1972).
156) a) Cazianis, C. T., Eaton, D. R.: Can. J. Chem. *52*, 2454 (1974);
b) Wilbur, D., Kreilick, R.: J. Chem. Phys. *52*, 1643 (1970);
c) Zelonka, R. A., Baird, M. C.: Chem. Commun. *1970*, 1448;
d) Brown, D. G., Maier, T., Drago, R. S.: Inorg. Chem. *10*, 2804 (1971).
157) a) Hoffman, B. M., Eames, T. B.: J. Am. Chem. Soc. *91*, 2169 (1969);
b) Brackman, W., Gaasbeck, C. J.: Rec. Trav. Chim. Pays-Bas *85*, 221 (1966);
c) Abakumov, G. A., Tikhonov, V. D.: Izv. Akad. Nauk SSSR, Ser. Khim. *1969*, 796; C. A. *71*, 22004 (1969).
158) Malatesta, U., Ingold, K. U.: J. Am. Chem. Soc. *95*, 6404 (1973).
159) Murata, Y., Mataga, N.: Bull. Chem. Soc. Jap. *44*, 354 (1971).
160) McConnell, A. A., MacNicol, D. D., Porte, A. L.: J. Chem. Soc. (A), *1971*, 3516.
161) Kabankin, A. S., Zhidomirov, G. M., Buchachenko, A. L.: J. Magn. Resonance *9*, 199 (1973).
162) Lozos, G. P., Hoffman, B. M.: J. Phys. Chem. *78*, 200 (1974).
163) a) Hogeveen, H., Gersmann, H. R., Praat, A. P.: Rec. Trav. Chim. Pays-Bas *86*, 1063 (1967);
b) Golubev, V. A., Zhdanov, R. I., Gida, V. M., Rozantsev, E. G.: Izv. Akad. Nauk SSSR, Ser. Khim. *1971*, 853; C. A. *75*, 63751 (1971).
164) a) Golubev, V. A., Zhdanov, R. I., Rozantsev, E. G.: Izv. Akad. Nauk SSSR, Ser. Khim. *1970*, 184; C. A. *72*, 111229 (1970);
b) Zhdanov, R. I., Golubev, V. A., Rozantsev, E. G.: Izv. Akad. Nauk SSSR, Ser. Khim. *1970*, 176; C. A. *72*, 111227 (1970).
165) a) Golubev, V. A., Voronina, G. N., Rozantsev, E. G.: Izv. Akad. Nauk SSSR, Ser. Khim. *1970*, 2605; C. A. *74*, 111841 (1971);
b) Golubev, V. A., Voronina, G. N., Rozantsev, E. G.: Izv. Akad. Nauk SSSR, Ser. Khim. *1972*, 161; C. A. *77*, 34321 (1972).
166) a) Golubev, V. A., Voronina, G. N.: Izv. Akad. Nauk SSSR, Ser. Khim. *1972*, 164; C. A. *77*, 34322 (1972);
b) Golubev, V. A., Sholle, V. D., Rozantsev, E. G.: Izv. Akad. Nauk SSSR, Ser. Khim. *1972*, 1202; C. A. *77*, 114165 (1972);
c) Sholle, V. D., Golubev, V. A., Rozantsev, E. G.: Izv. Akad. Nauk SSSR, Ser. Khim. *1972*, 1204; C. A. *77*, 114169 (1972).
167) Golubev, V. A., Zhdanov, R. I., Gida, V. M., Rozantsev, E. G.: Izv. Akad. Nauk SSSR, Ser. Khim. *1970*, 2815; C. A. *74*, 111325 (1971).
168) Golubev, V. A., Zhdanov, R. I., Protsishin, I. T., Rozantsev, E. G.: Izv. Akad. Nauk SSSR, Ser. Khim. *1970*, 2158; C. A. *74*, 76280 (1971).
169) Cadogan, J. I. G., Rowley, A. G.: JCS Chem. Commun. *1974*, 179.
170) Levenson, J. L., Kaplan, L.: JCS Chem. Commun. *1974*, 23.
171) Kornblum, N., Pinnick, H. W.: J. Org. Chem. *37*, 2050 (1972).
172) Zhdanov, R. I., Golubev, V. A., Gida, V. M., Rozantsev, E. G.: Izv. Akad. Nauk SSSR, Ser. Khim. *1970*, 2396; C. A. *74*, 141461 (1971).
173) Golubev, V. A.: Izv. Akad. Nauk SSSR, Ser. Khim. *1971*, 890; C. A. *75*, 48843 (1971).
174) Janzen, E. G. in: Topics in stereochemistry (eds. N. L. Allinger and E. L. Eliel), Vol. *6*, page 177. New York: Wiley-Interscience 1971.
175) Iwamura, N., Inamoto, N.: Bull. Chem. Soc. Jap. *43*, 860 (1970).
176) Minisci, F., Galli, R., Quilico, A.: Tetrahedron Letters *1963*, 785.
177) Gowenlock, B. G., Pfab, J., Kresze, G.: JCS Perkin II, *1974*, 511.
178) Aurich, H. G., Weiss, W.: Chem. Ber. *105*, 2389 (1972).
179) a) Weiss, W.: Diss. Univ. Marburg 1973;
b) Aurich, H. G., Weiss, W.: Liebigs Ann. Chem., submitted for publication.
180) Call, L., Ullman, E. F.: Tetrahedron Letters *1973*, 961.
181) Aurich, H. G., Kabs, K.: Angew. Chem. *82*, 634 (1970); Angew. Chem. internat. Ed. *9*, 636 (1970).
182) Maender, O. W., Janzen, E. G.: J. Org. Chem. *34*, 4072 (1969).
183) Craig, R. L., Roberts, J. S.: JCS Chem. Commun. *1972*, 1142.
184) Craig, R. L., Murray-Rust, J., Murray-Rust, P., Roberts, J. S.: JCS Chem. Commun. *1973*, 751.

185) a) Forrester, A. R., Ogilvy, M. M., Thomson, R. H.: J. Chem. Soc. (C), *1970*, 1081;
b) *loc. cit.*[5], page 215.
186) Perkins, M. J., Ward, P.: JCS Chem. Commun. *1973*, 883.
187) Keana, J. F. W., Baitis, F.: Tetrahedron Letters *1968*, 365.
188) Ullman, E. F., Call, L., Tseng, S. S.: J. Am. Chem. Soc. *95*, 1677 (1973).
189) Nelsen, S. F., Landis, II, R. T.: J. Am. Chem. Soc. *95*, 6454 (1973).
190) Balaban, A. T., Negoiţă, N., Baican, R.: J. Magn. Res. *9*, 1 (1973).
191) a) Razuvaev, G. A., Abakumov, G. A., Cherkasov, V. K.: Dokl. Akad. Nauk SSSR *198*, 601 (1971); Dokl. Chem. (Engl. Transl.) *198*, 439 (1971);
b) Abakumov, G. A., Cherkasov, V. K., Razuvaev, G. A.: Dokl. Akad. Nauk SSSR *197*, 823 (1971); Dokl. Chem. (Engl. Transl.) *197*, 265 (1971).
192) a) Grigo, U.: Diss. Univ. Marburg 1974;
b) Aurich, H. G., Grigo, U.: unpublished results.
193) Aurich, H. G., Weiss, W.: Chem. Ber. *106*, 2408 (1973).
194) a) Küttner, G.; Dipl.-Arbeit, Univ. Marburg 1974;
b) Aurich, H. G., Küttner, G.: unpublished results.

Received December 9, 1974

Cyclobutadienoids

Prof. Dr. K. Peter C. Vollhardt

Department of Chemistry, University of California, Berkeley 94720, U.S.A.

Contents

In recent years vigorous synthetic and theoretical research efforts have been directed at the preparation and investigation of cyclobutadiene, its derivatives and analogs. This has met with success in a number of areas related to the low temperature observation of the parent species itself, the synthesis of relatively stable substituted derivatives, and heterocyclic and annelated analogs, *e.g.*, benzocyclobutadiene and biphenylene derivatives. These "cyclobutadienoids", each class of which is of significant importance in its own right, have interesting physical and chemical characteristics due to incorporation of a potentially delocalized, antiaromatic four-membered ring. Recent progress in this area will be the topic of this report.

1. Introduction

The name "annulenes" has been given to completely conjugated, monocarbocyclic polyenes of general formula C_nH_n, the ring size being indicated by a number in brackets[1]. Cyclobutadiene ([4]annulene) and benzene ([6]annulene) are the smallest members of the series and the properties of these two molecules have long fascinated chemists.

Over one hundred years ago Kekulé proposed the now familiar hexagonal structure of benzene[2]. The subsequent development and wide acceptance of the electronic theory of valence enabled Armit and Robinson[3] to conclude that the unique properties of benzene, that is, its thermal stability, its tendency to undergo substitution rather than addition reactions, and the unusual behavior of some of its substituent groups, were connected with the presence of an "aromatic sextet" of electrons. The concept of the aromatic sextet was later brilliantly reformulated and extended by Hückel[4] when he published his famous $(4n + 2)$ π-electron rule, which stated that "Those monocyclic coplanar systems of trigonally hybrized atoms which contain $(4n + 2)$ π-electrons will possess relative electronic stability"[5]. The Hückel molecular orbital (HMO) theory[5] statisfactorily accounted for the stability of benzene, the 6 π-electrons entering bonding orbitals and forming a closed electronic shell. The HMO method as applied to the two nearest homologs of benzene, cyclobutadiene and cyclooctatetraene, using the geometrical formulation of Frost and Musulin[6], is shown in Fig. 1, together with the corresponding diagram for benzene. The difference in the pattern of orbitals between benzene, a $(4n + 2)$ π-system, and cyclobutadiene and cyclooctatetraene, two $(4n)$ π-systems, is readily apparent. The

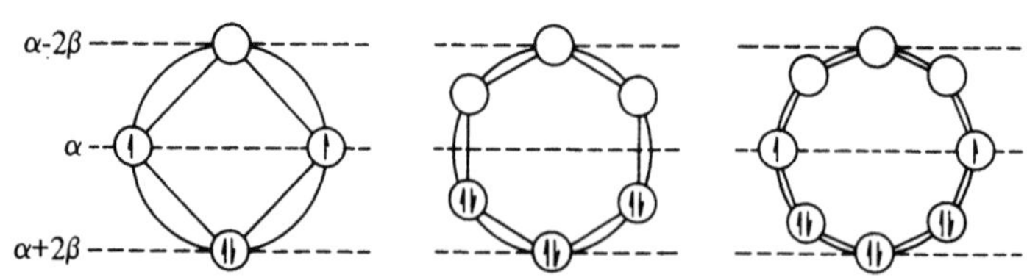

Fig. 1. HMO-diagrams for cyclobutadiene, benzene and cyclooctatetraene

two (4*n*) π-systems contain partially occupied non-bonding molecular orbitals (NBMO). There are two possible arrangements of the two electrons in the NBMOs; either the electrons are in one orbital with opposite spin, or each can occupy a separate orbital with parallel or antiparallel spins. Hund's Rule suggests that the latter arrangement with parallel spins will be of lower energy, but the difference in energy between the two states is likely to be small. The state with unpaired electrons will be a triplet diradical, whereas that with paired electrons will be a singlet.

Hückel recognized the difference between the [4*n*] and [4*n* + 2] series of annulenes and predicted aromatic properties for the latter, and non-aromatic properties for the former group. Subsequently, on the basis of experiment and also as the result of more sophisticated calculations, it was suggested that conjugation in the [4*n*]annulenes was destabilizing[7, 8]. The term antiaromatic was then suggested for these compounds. Experimental support for Hückel's rule included the synthesis of the cyclopentadienide anion and tropylium cation, as well as a series of [4*n* + 2]annulenes[1]. It was soon recognized that benzene, as the smallest member of the [4*n* + 2] aromatic series, possessed rather special properties, due to the absence of ring strain and nonbonded interactions. Cyclobutadiene[1], the smallest member of the [4*n*] series, also has rather special antiaromatic properties, due to its inability to distort from planarity to remove the destabilizing effect of conjugation. Thus, cyclobutadiene is an extremely reactive molecule, which has only been observed using rather special techniques.

1

However, stabilization of the cyclobutadiene nucleus can be achieved by a variety of methods (substitution, heteroatom incorporation, complexation, annelation) leading to cyclobutadienoid structures containing completely unsaturated four-membered rings. Generally, the effect of incorporating a cyclobutadiene nucleus on the electronic structure of a particular system will be the generation of relatively high-lying highest occupied molecular orbitals (HOMOs) and low-lying lowest unoccupied molecular orbitals (LUMOs). This in turn will lead to relatively low energy transitions in the electronic spectra and an increased paramagnetic ring current contribution to the overall ring current of the π-system[1, 6, 16] due to the four-membered ring. In the following account, the various ways by which cyclobutadiene might be and has been stabilized, will be discussed. The physical properties, particularly the electronic and NMR spectra, and chemical behavior of the resulting molecules will be described as a measure of residual cyclobutadienoid character in the modified cyclobutadiene.

2. Cyclobutadienes

2.1. The Parent System

Several exhaustive reviews have dealt with synthetic and theoretical aspects of the cyclobutadiene problem[9]. The usual routes for the *in situ* generation of cyclobutadienes have involved in the last step one of the following reactions: dehalogenation or dehydrohalogenation of substituted cyclobutanes or cyclobutenes, acetylene dimerizations, ring closure reactions, Hofmann elimination of bis(dimethylamino) cyclobutane dimethiodides and retro-Diels-Alder reactions. Recently, applying low temperature photolysis techniques, cyclobutadiene was generated[10b] in a solvent- or solid gas-matrix at 8–20 K and a review has appeared[10a]. The results of this research can be briefly summarized as follows. Cyclobutadiene generated by the matrix isolation procedure is square planar, a geometry that has been assigned to the triplet species by theoretical calculations. On the other hand, chemical evidence seems to point to a singlet ground state of cyclobutadiene, since, once generated, it dimerizes sterospecifically. However, since the singlet-triplet energy gap has been estimated to be only a few kcal/mole and the stereospecificity of cycloaddition reactions of a triplet cyclobutadiene is unknown, the ground state multiplicity of *1* is still a matter of speculation. The electronic spectrum of cyclobutadiene exhibits a relatively high wavelength band at 300–305 nm, a position which does not seem to correlate well with the calculated absorption spectra of either singlet or triplet cyclobutadiene. Experiments with moderately substituted cyclobutadienes have revealed the existence of a valence isomeric equilibrium *1a* ⇌ *1b*, with an estimated barrier for double bond shift of ca. 8 kcal/mole.

1a *1b*

Our knowledge about the parent cyclobutadiene system and some simple substituted derivatives has thus improved greatly over the last few years, although a number of unsolved problems still remain, challenging the ingenuity of able experimentators.

For the preparative chemist the question arises of how to stabilize a fascinating molecule like cyclobutadiene, such that relatively non-labile substances are obtained which then could be further subjected to the investigator's scrutiny. The physical and chemical properties of such stabilized cyclobutadienes are of great interest, not only because their determination would enable one to draw conclusions about the nature of the parent species, but also because in electronically stabilized cyclobutadienes the electronic structure of the original system might be distorted to such an extent that new species, worthy of investigation in their own right, are produced. These species might be arbitrarily designated "cyclobutadienoids" in as much as they retain certain characteristics of the cyclobutadiene nucleus, in addition to the development of their own distinct patterns of physical and chemical behavior. Since significant distortion of the cyclobutadiene nucleus occurs even on simple substitution, such systems may be included in the cyclobutadienoid classification.

2.2. Substituted Cyclobutadienes

One way to stabilize the cyclobutadiene nucleus is by substitution with bulky substituents. Although early attempts in this area[9, 11] did not lead to any isolable cyclobutadiene species, recently Masamune[12], Krebs[13], Maier[14] and their respective co-workers were able to synthesize stable cyclobutadienes bearing bulky groups. Krebs' group[13] generated a palladium dichloride complex of *3* via reaction of thiacycloheptin *2* with $(C_6H_5CN)_2PdCl_2$. Cyclobutadiene *3* was then liberated by the action of ethylene-bis (diphenylphosphane) on the complex. Compound *3* is

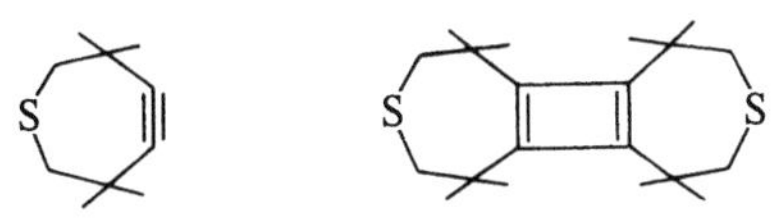

2 *3*

remarkably thermally stable (m.p. 240 °C!), but oxygen sensitive. The electronic spectrum of the yellow material shows only one band at 227 nm tailing into the visible region. Strong indication that *3* is a cyclically delocalized 4π-electron system is obtained from the photoelectron spectrum[15] which reveals an unusually high lying highest occupied molecular orbital as expected for an antiaromatic molecule[16]. An X-ray analysis[94] depicts the four ring as a planar rectangle with two long and two short bonds as indicated in structure *3*.

Masamune *et al.*,[12] and Maier *et al.*,[14] were able to synthesize the unstable bright yellow tri-tert.-butylcyclobutadiene *4*

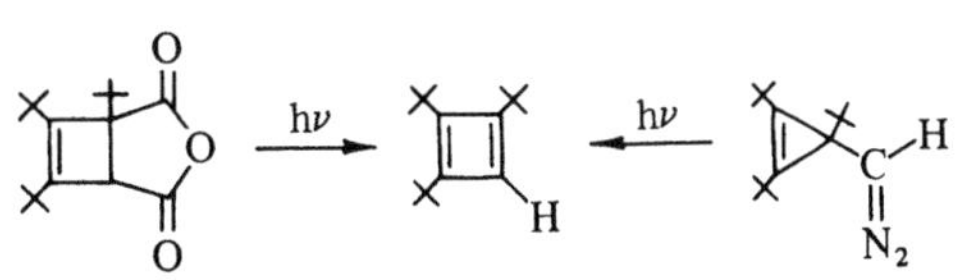

5 *4* *6*

from the cyclopropenyldiazomethane derivative *6* and the cyclobutenedicarboxylic anhydride *5*, respectively. The chemical shift observed for the ring proton in *4* (τ 4.62, THF-d8) points to the paratropic[6] character of *4* resulting in a relatively high field absorption. Compound *4* dimerizes rapidly at room temperature and is extremely air sensitive. Replacement of the ringproton in *4* with a carbomethoxy-

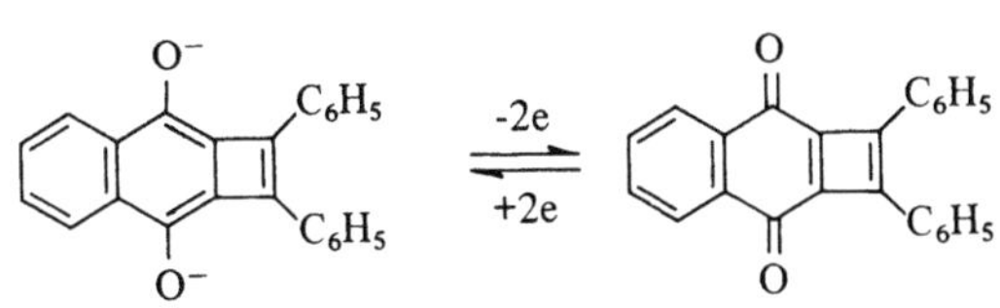

7 *8*

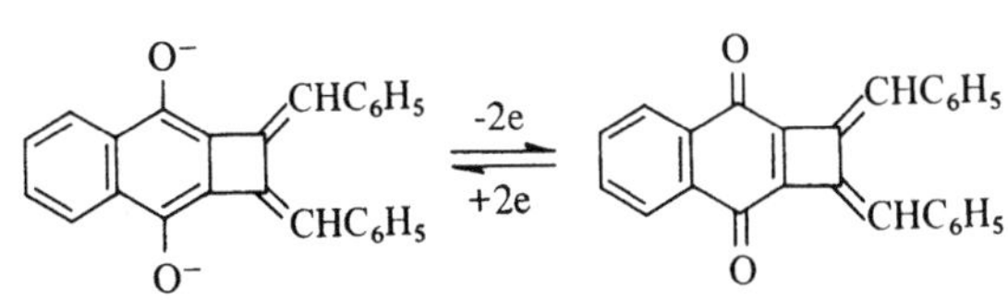

9 *10*

group stabilizes the 4π-system further and the resulting derivative is thermally stable (m.p. 70 °C), but still very oxygen sensitive[12].

These compounds have to be regarded as the closest analogs to the parent cyclobutadiene isolated to date and their preparation indicates the powerful stabilizing ability of sterically demanding groups.

Electrochemical evidence for the antiaromaticity of cyclobutadiene has been provided by Breslow and co-workers[17]. The oxidation potentials for the hydroquinone dianion *7* (−1.50 V, −0.68 V; versus Ag – AgCl, at Pt electrode) are substantially more negative than the oxidation potentials of model *9* (−1.22 V, −0.45 V). In the *7* → *8* conversion a dimethylenecyclobutene derivative, with only a small degree of possible cyclobutadiene character, is converted into a full cyclobutadiene, presumably partially stabilized by the ketofunctions. The data indicate that the cyclobutadiene resonance ***destabilization*** amounts to at least 12 kcal/mole, and an estimate of the true antiaromatic destabilization energy of 15–20 kcal/mole has been made[17].

2.3. Push- Pull-, Hetero- and Other Stabilized Cyclobutadienes

Stabilization of cyclobutadienes may be achieved, if the cyclobutadiene structure is only *one* of the important resonance forms. This is the case in the so-called "push-pull" cyclobutadienes.

Simple MO-calculations[18, 19] show that cyclobutadienes substituted with conjugated electron-donor ("push") and electron-acceptor ("pull") groups have higher electron delocalization energies than cyclobutadiene itself. Furthermore, the degeneracy of the two nonbonding MO's is removed, thus giving rise to square singlet ground states[18, 20].

It has been predicted[21–23] that tetra-substitution of cyclobutadiene with donor (D) and acceptor (A) groups, as in *11*, will produce an optimal stabilizing effect, due to contributing forms *11c* of tetramethylenecyclobutane structure.

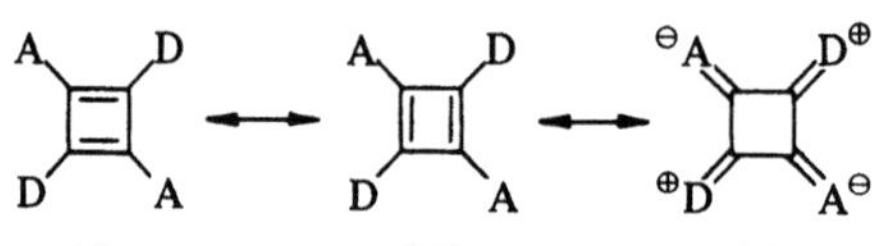

11a *11b* *11c*

Indeed, push-pull cyclobutadienes of the type *12* are stable isolable substances, with seemingly strong contributions of canonical structure *11c* to the ground-state[21, 22]. An X-ray structural determination of *12* (R = CH_3CH_2, R′ =

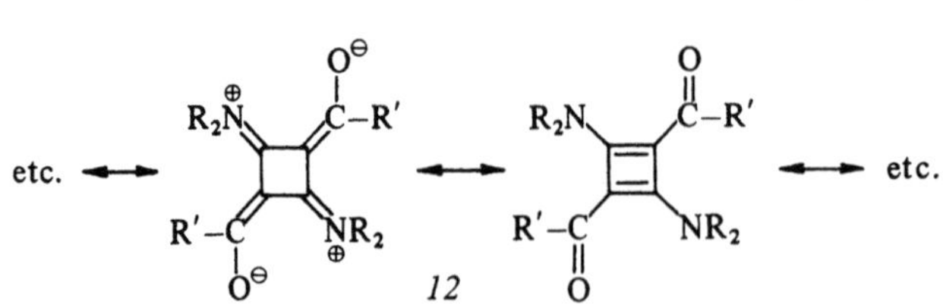

12

$CH_3CH_2O)$[24] revealed the molecule to be of deformed square planar geometry with significant double bond character in the C–N-bonds. In contrast to an earlier report[25a], the photoelectron spectrum shows it to exist as two strongly interacting aminoacryl ester moieties and the available experimental and theoretical data point to *12* as being a delocalized [4]annulene[25b].

Substitution of one of the CH – units of cyclobutadiene with nitrogen leads to azacyclobutadiene, which is predicted to gain significant stabilization from donor groups in the 3-position[26], *e.g.*, *13*. The first nonannelated derivative of such a system has recently been synthesized[27] by the flash thermolysis of 4,5,6-tris(dimethylamino)-1,2,3-triazin. A red compound was obtained to which, on the basis of its physical properties, was assigned structure *14*. Azacyclobutadiene *14* is stable at low temperatures and has a complicated chemical behavior. Its electronic spectrum shows a strong band at 527 nm ($\epsilon \approx 20000$) giving rise to the observed color. Resonance contributor *13b* seems to play an important role in the description of the ground state of *14*, enabling the isolation of this interesting species[97].

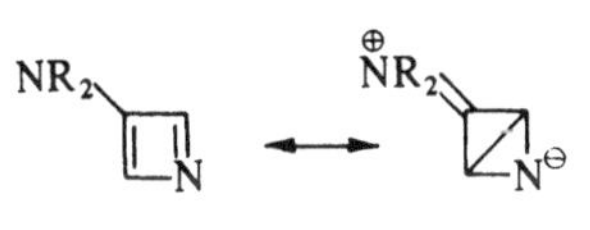

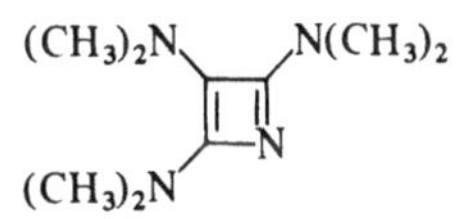

13a *13b* *14*

An alternative way of stabilizing the cyclobutadiene nucleus is by complexation to transition metals. This was predicted in 1956[28], and experimentally verified by the isolation of tetramethylcyclobutadiene nickel chloride[29]. Since then, the chemistry of cyclobutadiene complexes has become a major research area and several reviews have appeared[30]. Particularly, Pettit's cyclobutadiene iron tricarbonyl *15* is a useful starting material for the *in situ* generation of cyclobutadiene by oxidative degradation[31]. Due to the particular bonding characteristics of these metal com-

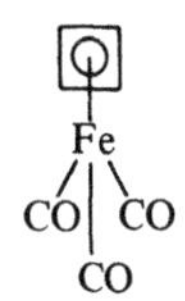

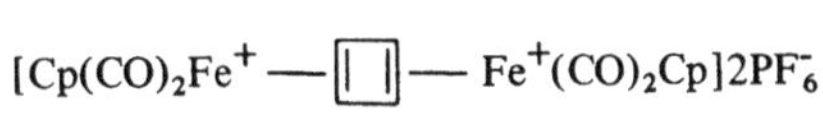

15 *16* Cp=η^5-cyclopentadienyl

plexes[32], involving "forward" donation of electron density from the HOMO of the diene unit to an empty *d*-orbital of appropriate symmetry, and "back" donation of electron density from a filled *d*-level to the LUMO of the cyclobutadiene, the electronic structure of the four ring is considerably perturbed. This is borne out by the physical and chemical properties of cyclobutadiene transition metal complexes, which reveal very little similarity to the properties of the free ligand itself. Interestingly, a η^2-cyclobutadienoid iron complex *16* has recently been reported which also shows remarkable stability[33].

A rather intriguing compound is 3,4-dimethylenecyclobutene *17*, which can be viewed as a cyclobutadienoid structure by virtue of some degree of delocalization

17a *17b*

across the central bond of the exocyclic diene unit. Indeed, the resonance energy per π-electron in *17* is -0.028β, classifying *17* as an antiaromatic moiety[34]. The fact that *17* can be isolated as a relatively stable molecule was suggested to be a reflection of the unfavorable transition state energies for Diels Alder type dimerization of *17*, which would lead to substituted cyclobutadiene structures.

3. Benzocyclobutadienes

3.1. Benzocyclobutadiene and Substituted Derivatives

Another method of stabilizing the cyclobutadiene ring might be by annelation. The most significant resonance contributor of benzocyclobutadiene is expected to be *18c*, in which the four-membered ring is not in the form of a cyclobutadiene. However, one might argue that bond fixation as in *18c* will result in the loss of the aromatic character of the benzene ring. Thus, like the parent [4]annulene, benzocyclobutadiene is expected to be an extremely reactive molecule and has not yet been isolated[9a]. Its existence as a transient intermediate has been postulated in

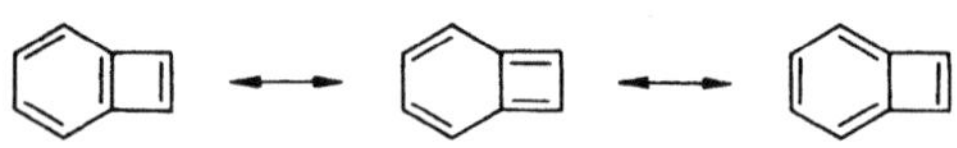

18a *18b* *18c*

many reactions. Benzocyclobutadienes have been generated in ways analogous to those for cyclobutadiene, that is, via halogenated benzocyclobutenes or cycloadditions of benzyne to acetylenes. The parent compound *18*, formed *in situ,* dimerizes either angularly to *19* or linearly to *20*, depending on reaction conditions. Attempts to reduce the reactivity of *18* by blocking the 1,2-positions with suitable groups have, until recently, been unsuccessful. Thus, attempted generation of 1-methyl-2-phenyl-benzocyclobutadiene[35], 4,6-dimethyl-1-*t*-butylbenzocyclobutadiene[36], 1-*t*-butyl- and 1,2-di-*t*-butylbenzocyclobutadiene[37] gave only dimeric products.

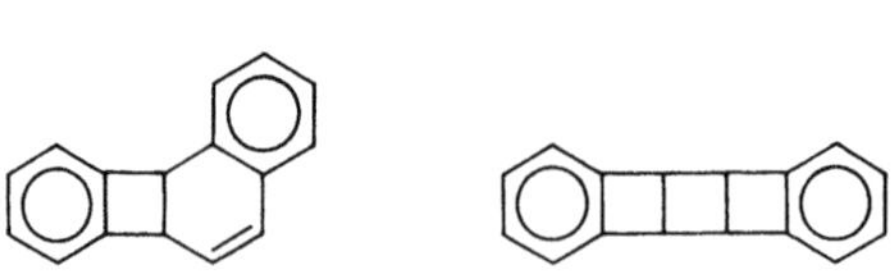

19 *20*

On the other hand, heating *trans*-bisacetylene *21* to 110° for several minutes generated the red 3,4,5,6-tetramethyl-1,2-diphenylbenzocyclobutadiene *22* in 20%

yield[38]. Crystalline *22* is transformed directly into the dimeric dibenzocyclooctatetraene *23* on heating above 80°. The methyl absorptions in the NMR-spectrum of *22* appear at relatively high field (τ 8.07), possibly a reflection of the potentially paratropic character of the $4n$-π cycle. The electronic spectrum exhibits bands up to 468 nm, a typical characteristic of planar 8π-electron systems (vide infra). Unfortunately, *22* does not bear any ring protons. An estimate of the antiaromaticity of *22* by the NMR ring current criterion[6] is therefore very difficult.

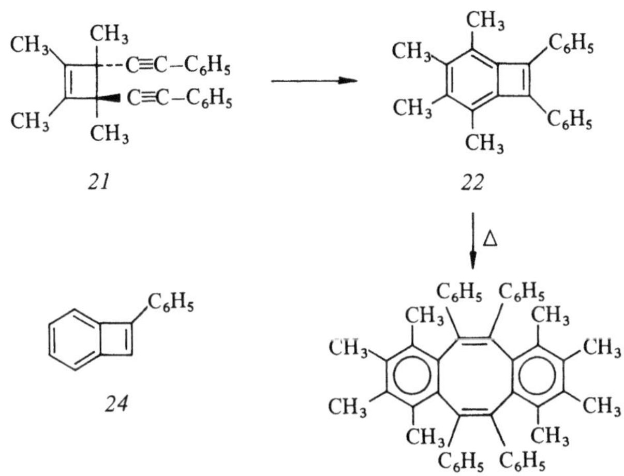

21 *22* *24* *23*

Ota *et al.*[39] have recently observed an intermediate in the matrix irradiation of diphenylacetylene, to which they assigned the 1-phenylbenzocyclobutadiene structure *24* on the basis of a mass determination and a UV-spectrum. No NMR data were reported, however, and a chemical structural proof would be desirable.

3.2. Other Benzocyclobutadiene Derivatives

Vapor flash pyrolysis of 4-phenylbenzo-1,2,3-triazine *25* led, amongst other products, to the red 2-phenyl-1-azabenzocyclobutadiene *26*, stable at –80 °C[40]. On

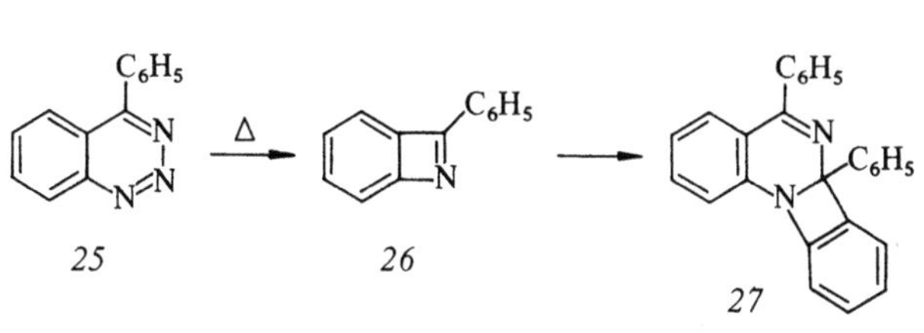

25 *26* *27*

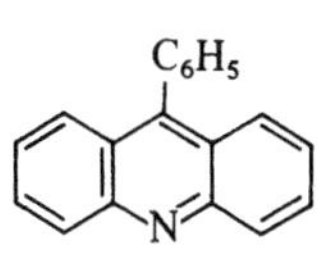

28

warming dimerization occured to an angular dimer (possibly *27*) which at higher temperatures eliminated benzonitrile to give 9-phenylacridine *28*. Cycloadditions to the imine double bond were also observed, and compound *26* seems to behave very much like the carbocyclic analog in its reactions. No spectral data on *26* have been reported, presumably due to the fact that *26* after pyrolytical generation is contaminated with side products.

Push-pull stabilized benzocyclobutadienes are not known, but several benzocyclobutadienes are isolable as stable transition metal complexes[9], notably *29*, synthesized by Pettit and his co-workers[41]. The diene character of the uncomplexed

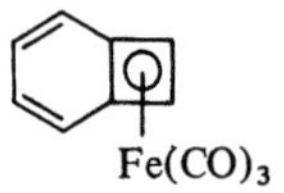

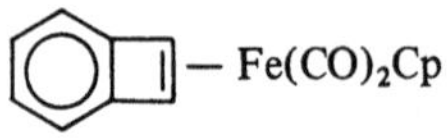

29 *30* Cp=η^5-cyclopentadienyl

benzenic double bonds is revealed on exposure of *29* to $Fe(CO)_5$ and light, leading to benzocyclobutadiene-bis-iron tricabonyl[42]. Similar to cyclobutadiene complex *16*, a η^2-benzocyclobutadiene complex *30* has been shown to be synthetically accessible[43]. Compound *30* is, however, unstable at room temperature and on exposure to oxygen, and the labile ligand is easily trapped in the presence of reactive dienes.

3.3. Nonbenzenic Analogs of Benzocyclobutadiene

A simple explanation for the reactivity of benzocyclobutadiene may be found in the desire of benzene to delocalize, thus leading to increased cyclobutadienoid character and therefore reactivity in the four ring. One would therefore expect that replacement of the benzene ring by another less or more delocalized π-cycle should have a corresponding effect on stability and reactivity of the resulting benzocyclobutadiene analog. Several compounds of this type have been prepared and reveal fascinating properties.

Cyclobutadienocyclopentadienyl anion *33* has been generated by treatment of the chlorodiene *31* with base, presumably via the intermediate formation of triene *32*[44]. As strongest evidence for the formation of *33* can be regarded the fact that

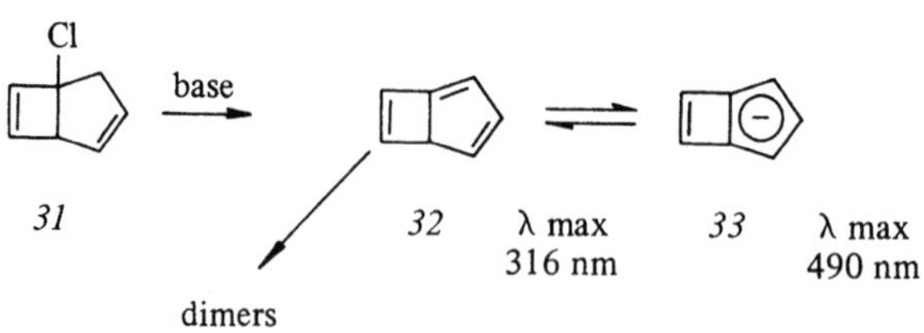

generation of *33* in protic deuterated solvents leads to dimers of *32* with two incorporated deuterium atoms. Anion *33*, characterized by a high wave length electronic absorption band (490 nm)[45], is very reactive and protonates only at C–1 to give *32*. The pKa of *33* is raised by 11 units relative to the pKa of the cyclopentadienyl anion[46]. Thus, cyclopentadienyl anion is destabilized by ca. 15 kcal/mole by fusion to cyclobutadiene. This destabilization is presumably partially a reflection of strain, but more importantly an indication of the destabilizing effect of antiaromatic cyclobutadienoid structures[47].

Attempted generation of anion *33* from the relatively stable triene *37* under a variety of conditions led to dimers derived from the isomeric triene *32*[48]. However, *33* could be formed in the gas phase, and its proton affinity was considerably larger than that of the cyclopentadienyl anion.

Since the degree of delocalization and "aromaticity" in the cyclopentadienyl anion could be considered to be approximately equal to that in benzene[6], it is not surprising that the synthetic difficulties in the preparation of cyclobutadienocyclopentadienyl anion *33* resemble those encountered in the attempted isolation of benzocyclobutadiene. Fusion of less aromatic π-cycles to cyclobutadiene, however, should lead to relatively stable and isolable benzocyclobutadiene analogs.

In analogy to the preparation of bicyclic triene *37* from bisacetylene *34*[49], the diethynyloxiranes *35* and thiiranes *36* (both *cis* and *trans* isomers can be employed; *trans*- *36* leads to desulfurized product only) were subjected to gas phase flow-pyrolyses to effect clean, though low yielding, rearrangement to furo- and thienocyclobutadiene *38*[50] and *39*[51], respectively, systems that are isoelectronic with anion *33*. Both *38* and *39* are planar 8π-systems and as such exhibit typical

34	X = ·CH_2	*37*	
35	X = O	*38*	*40*
36	X = S	*39*	*41*

antiaromatic properties. The compounds are extremely air sensitive and dimerize linearly even in dilute solutions to difuro- and dithienocyclooctatetraene, respectively. The cyclobutene double bond is very reactive and can be added to dienes or be selectively reduced. Comparison of the physical properties of *38* and *39* with those of the corresponding cyclobutane model compounds *40*[52] and *41*[53] reveals the drastic effect of introducing an additional double bond into the cyclobutahetero [5]annulene nucleus. Thus, *40* and *41*, in contrast to *38* and *39*, are relatively air and thermally stable. The chemical shift of the furan proton[54] in *38* is at considerably higher field (Δδ = 0.77 p.p.m.) than the chemical shift of the corresponding proton in *40*, indicating a significant paratropic component in furocyclobutadiene. In addition, one can observe a dramatic change in the UV-spectrum when going from *40* (endabsorption only) to *38* (highly complex spectrum with absorption up to 387 nm) in accord with theoretical expectations for a planar 4*n* π-system[16]. Similar observations can be made in the comparison of *39* and *41*[51].

A crystalline 1,2-diphenylderivative *44* of thienocyclobutadiene *39* has been prepared by the bis-Wittig[55] reaction of 3,4-diphenylcyclobutadiene quinone *42*

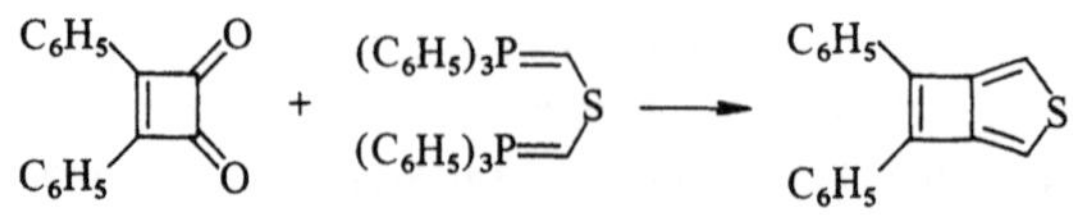

42 43 44

with the bisylid 43[56]. The spectral data on 44 resemble those on the parent system indicating that the two phenyl groups exert stabilizing influence by virtue of their steric bulk only. The chemical behavior of 44 is very similar to the chemical behavior of tetramethyldiphenylbenzocyclobutadiene 22 (Section 3.1.), including the linear cycloaddition to tetracyanoethylene which leads to the rearranged adduct 45[57].

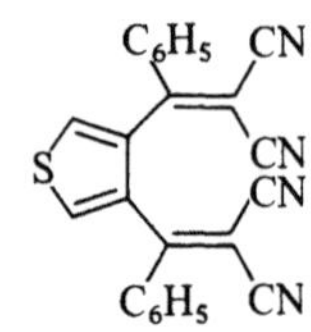

45

Several other nonbenzenic formal analogs of benzocyclobutadiene have become known in recent years but their physical and chemical properties are not well understood. Cyclobutadienocyclopentadienone 46, a bicyclotropone, for instance has a trienetype chromophore (λ max 284 nm) and polymerizes in moderately concentrated solutions even in the cold[58]. No well defined products can be isolated derived from 46. Some higher vinyloges of 46 (*e.g.*, 47) have been obtained as stable iron tricarbonyl complexes and do not display cyclobutadienoid character[59].

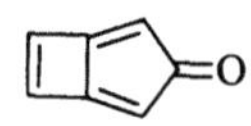

46

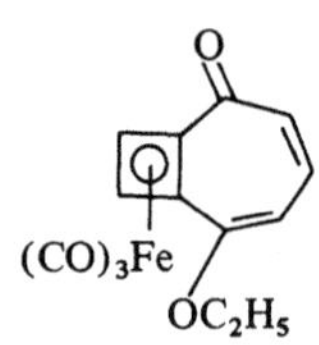

47

Fusion of cyclobutadiene to another $4n$ π-system leads to a π-cycle containing a total number of $(4n + 2)$ π-electrons, a formally aromatic arrangement. Both 48[60] and 49[61] (the latter in the charge-separated form 49b) could be viewed as planar examples of this class of compounds. Although rather reactive, both 48 and 49 exhibit NMR spectra with unusually low chemical shifts for the cyclobutenyl and

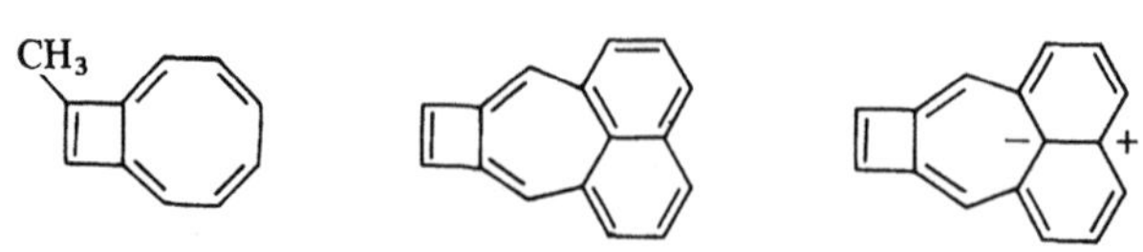

48 49a 49b

other protons. This observation could be explained by the assumed occurrence of an induced diamagnetic ringcurrent when *48* and *49* are exposed to a magnetic field. Further work is in order to enable a better understanding of the π-electronic characteristics of such compounds.

3.4. Higher Aromatic Analogs of Benzocyclobutadiene

Benzocyclobutadiene *67* is the simplest member of a series of completely unsaturated, polynuclear hydrocarbons derived by fusion of a benzenoid system with one side of a cyclobutadiene ring. These higher aromatic analogs of benzocyclobutadiene have been the subject of a number of theoretical calculations[18, 62]. Dewar[63] calculated the difference in resonance energy ΔE_R between the various cyclobutadiene derivatives and the pair of π-systems obtained by breaking the two 'single' four-ring bonds. This value of ΔE_R was used as an indication of the degree of antiaromaticity induced by the four-ring in the respective compounds. Molecules with large negative ΔE_R values should be chemically unstable, tending to undergo reactions which remove the unfavorable effect of the destabilizing conjugation through the four-membered ring. The known stabilities of a group of such compounds (or their derivatives) seem to correlate well with the calculated values (Table 1). In addition, a correlation was detected between the calculated average bond lengths of the four-

Table 1. ΔE_R values (see text) for some benzenoid benzocyclobutadiene analogs

Structure	ΔE_R (eV)
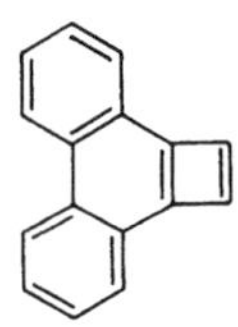	−0.716
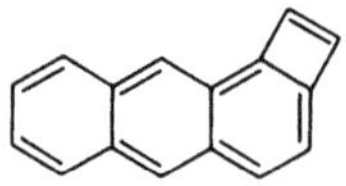	−0.656
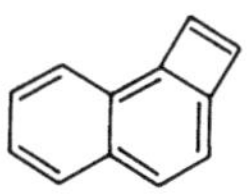	−0.615
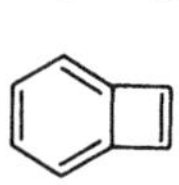	−0.488
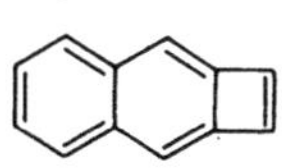	−0.314
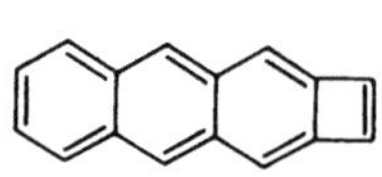	−0.243

ring and the overall stability of the system; the larger the average bond length, the more stable should the system be[62, 63].

None of the cyclobutadienoid compounds depicted in Table 1 has been isolated, although some substituted derivatives are known and reviews have appeared[9a, 64]. Substitution of the four-ring seems a necessary prerequisite for stabilization, thus, naphthocyclobutadienes *50* and *51*, and anthrocyclobutadiene *52* have been obtained as relatively stable substances. The diphenylderivative *50* is a bright red com-

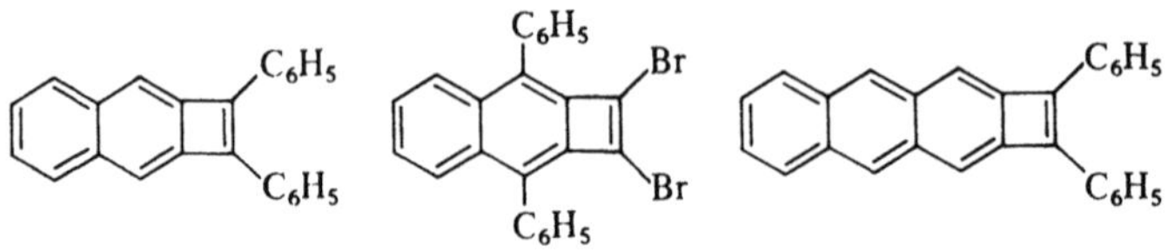

50 *51* *52*

pound (λ max up to 455 nm), and the paramagnetic ringcurrent contribution of the cyclobutadienoid ring leads to an unusually high field absorption (singlet, τ 3.50) of the benzenic protons nearest to the four-ring. Interestingly, the dibromo-diphenyl derivative *51* shows electronic absorptions at much lower wavelength, indicating a pronounced effect of the bromine atoms on the electronic structure of the system. All three compounds *50*–*52* showed, like benzocyclobutadiene, a high reactivity of the 1,2-double bond. Generation of less substituted derivatives of *50* in the presence of reactive dienes leads to the successful isolation of Diels-Alder adducts.

Recently phenanthrocyclobutadiene *53* has been generated from its iron tricarbonyl complex and trapped with dienes and tetracyanoethylene, respectively[65]. This finding reopens the question of the ground state spin multiplicity in *53*[66], a molecule of considerable cyclobutadienoid character (see Table 1).

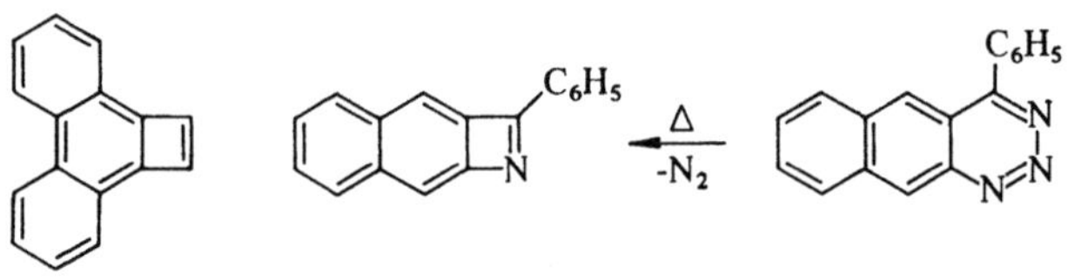

53 *54* *55*

Very little is known about heterocyclic higher analogs of benzocyclobutadiene, although naphthoderivative *54* could be generated by pyrolytical nitrogen exclusion from 4-phenyl-1,2,3-naphtho[2,3-*d*]triazine *55*[40]. Further activity in this area is to be expected.

4. Biphenylenes

4.1. Biphenylene and Benzobiphenylenes

Biphenylene *56* is formally the dibenzo derivative of cyclobutadiene. However, because of its unusual physical and chemical properties, it is more accurately viewed as the parent member of a unique series of polynuclear aromatic hydrocarbons. Several exhaustive reviews have described the chemistry of biphenylene and related compounds[9, 67, 68].

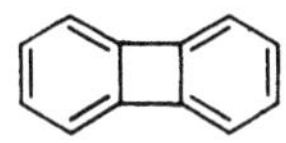

56

The biphenylene system has interested theoretical chemists for over 20 years. LCAO-MO calculations[9a, c, 18, 62, 63, 69] predicted that the four-membered ring is closer in structure to tetramethylenecyclobutane than to cyclobutadiene. Structural analyses[70] have supported this prediction and have shown *56* to be completely planar with the bond-lengths and angles shown in Fig. 2. Evaluation of the proton NMR spectrum of biphenylene indicated that the partial bond fixation shown by

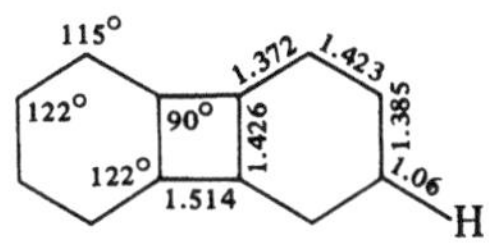

Fig. 2. X-ray structure of biphenylene (Å)

X-ray analysis is also present in biphenylene when in solution[71]. However, Binsch[72] found that *56* showed neither strong first-order nor second-order bond fixation and defined it as aromatic, as opposed to benzocyclobutadiene.

Although the two benzene rings in *56* are linked by almost 'single' bonds, there is ample evidence to demonstrate that they are not independent and that substantial delocalization is present over the entire π-system. Thus, the electronic spectrum has two sets of hands at 235–260 and 330–370 nm, significantly different from biphenyl which has only one absorption band at 250 nm. A charge-transfer complex of *56* with tetracyanoethylene, presumably of structure *57*, was shown to be more stable than the corresponding fluorene complex[73]. The infra-red absorptions for

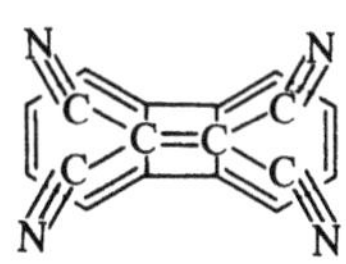

57

the cyano-groups in cyano-biphenylenes are shifted to lower frequency when compared with the corresponding cyano-benzenoid system obtained by removing the 4-membered ring, again supporting the view that π-delocalization occurs across the four-ring[9a].

Comparison of the calculated[63] and experimentally derived[74] heats of formation of biphenylene, suggest a destabilization of 74.2 kcal in *56*. This has usually been associated with the strain energy contained in the four-membered ring. To what extent the cyclobutadienoid character of this ring contributes to the observed discrepancy between experimental and calculated energy content is a matter of speculation.

The molar susceptibility of biphenylene, which is less than that of only one benzene ring, and the negligible value for the diamagnetic anisotropy, can be rationalized in terms of an induced paramagnetic ring-current in the cyclobutadiene ring cancelling the diamagnetic ring-current effect of the two benzene-rings.

The possible antiaromaticity of the four-membered ring in biphenylene and benzobiphenylenes has been investigated by Figeys[9c)] according to the criterion based on the magnetic properties of these compounds, in particular the proton NMR data. In contrast to polycyclic benzenoid systems the NMR signals of the protons of biphenylene appear at higher field than the benzene resonance, and the same is true of certain protons in benzobiphenylenes. Table 2 shows ring-currents, calculated by the McWeeny method, and observed chemical shifts for benzene, biphenylene and selected benzobiphenylenes. The observed upfield shifts might be due either to

Table 2. Calculated ringcurrent intensities and chemical shifts in benzene, biphenylene, benzo[*a*]biphenylene, benzo[*b*]biphenylene and dibenzo[*a, c*]biphenylene[9 c)]

Compound	Ring	Ring-current intensity	Proton	Chemical shift (τ, $CDCl_3$)
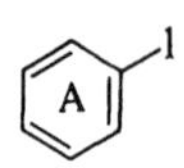	A	1.000	1	2.75
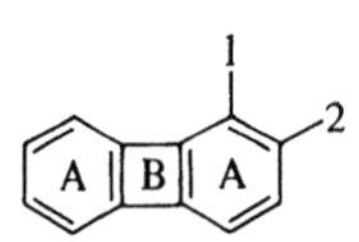	A	0.563	1	3.402
	B	−1.028	2	3.298
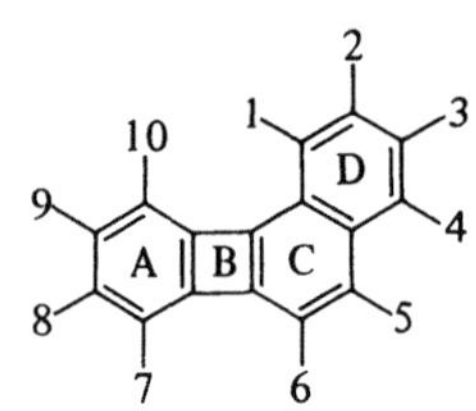	A	0.374	1−5	2.750
	B	−1.474	6	3.117
	C	0.705	7−10	3.533
	D	0.875		
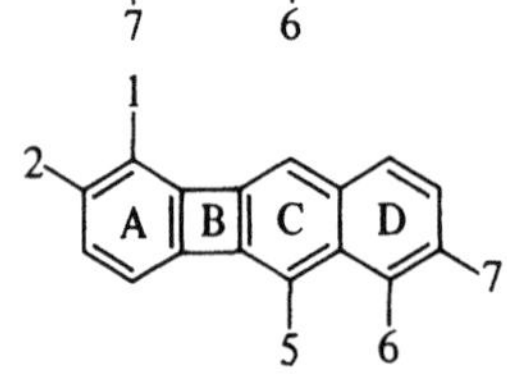	A	0.578	1,2,5	3.093
	B	−0.949	6	2.573
	C	0.577	7	2.774
	D	0.929		
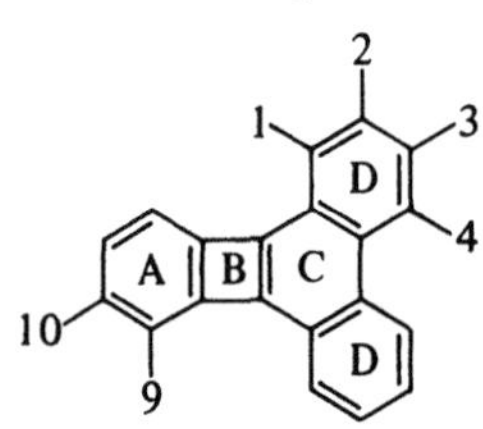	A	0.285	1−3	2.400
	B	−1.668	4	1.500
	C	0.720	9−10	3.500
	D	0.995		

the occurrence of very low deshielding ring-currents in the various benzenic rings, or to the presence of an induced paramagnetic shielding ring-current in the four-membered ring. Evidence for the second hypothesis is provided by inspection of the NMR spectrum of biphenylene and benzo[*b*]biphenylene[75], where, if the first hypothesis were true, H_1 should resonate at lower field than H_2, whatever the relative values of the individual ring-currents, owing to the closer proximity of the diamagnetic effect of the benzenic rings A and C, D respectively (Table 2). In fact, however, H_1 appears at similar or higher field compared with H_2. Further evidence supporting this hypothesis is found in the calculations which indicate strong paramagnetic ring-currents in ring B.

4.2. Carbocyclic Analogs of Biphenylene

Replacement of one benzene ring in biphenylene by another $(4n + 2)$ π-system gives rise to an interesting series of cyclobutadienoids, the relative paratropic character of which should vary with the degree of delocalization associated with the new ring (Table 3).

Table 3. Spectral data on some carbocyclic biphenylene analogs

Compound	NMR chemical lifts (solvent, τ, order as indicated in the structure)	Electronic spectra (λ max, solvent, nm)	Ref.
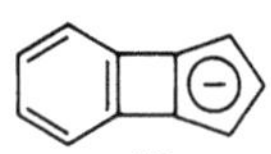 58	–	$(CH_3)_2SO$; 500 sh, 590	76)
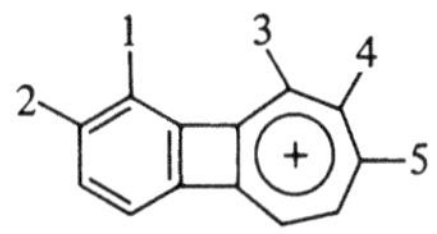 59	CH_3CN; 2.78, 2.78 2.20–2.52, 1.80, 2.20–2.52	CH_3CN; 226, 233, 279, 328, 388, 399, 510.	77)
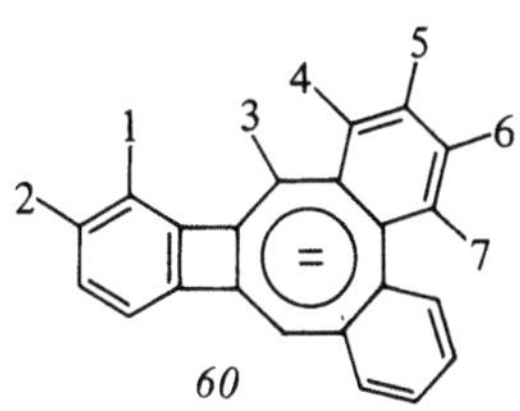 60	THF-d_8; 4.87, 4.51, 4.23, 2.52, 3.90, 3.90, 2.52.	$(CH_3O)_2(CH_2)_2$; 380, 486, 625.	78)
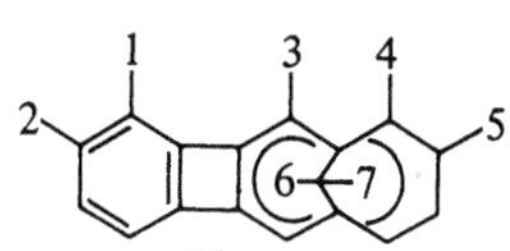 61	CCl_4; 3.27, 3.27, 3.32, 3.39, 3.06, 7.55, 9.71.	CH_3CH_2OH; 223, 280, 320, 388, 398.	79)

The chemical behavior of the norbiphenylene anion *58* resembles that of the cyclobutadienocyclopentadienyl anion *33*. Thus, *58* gives a mixture of dimers on protonation[76]. No NMR-spectrum of *58* has been reported. The spectral data on *59*[77] and *60*[78] indicate considerable conjugative interaction throughout the respective systems. In addition, paramagnetic ring-current contributions arising from the cyclobutadienoid character of the four-ring lead to unusual highfield shifts of the protons close to this ring in the NMR spectrum. This is particularly striking in *60* where chemical shift differences of up to 2.35 ppm are observed between formally analogous benzenic protons. In the neutral compound *61* the paratropicity of the four-ring seems diminished and only slight upfield shifts for $H_{1,2,3}$ are observed[79]. The large chemical shift difference for the diastereotopic bridge protons seems to be largely due to the divinylhomobenzene character of *61, e.g., 62*, as evidenced by ^{13}C-NMR spectroscopy[80], although the low chemical shift of H_6 might be partially

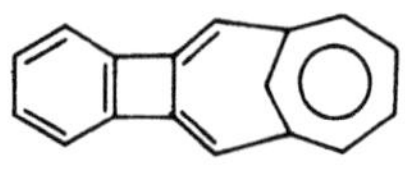

62

due to its location inside the deshielding cone of the paramagnetic ring-current in the four-ring[79].

It should also be mentioned that the anion corresponding to *59* seems to exhibit significant diatropic character, although no specific data have been published[96].

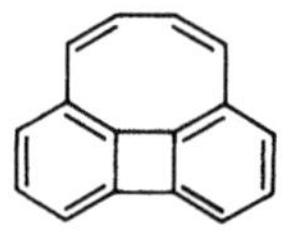

63

Another curious biphenylene system, compound *63*, has been reported recently[93]. The chemical shifts of the biphenylene (τ 3.6–4.4) and cyclooctatetraene protons (τ 5.38) indicate significant paramagnetic contributions from the cyclobutadienoid and the planar 8π-ring to the overall ringcurrent in the molecule. Compound *63* also exhibits electronic absorptions at unusually high wavelenghts (621 nm).

4.3. Heterocyclic Analogs of Biphenylene

Parallel to the carbocyclic series of biphenylene analogs, there is also a heterocyclic series, which includes systems in which one or both benzene rings of biphenylene have been replaced by heteroannulenes of the pyridine and pyrrole type[6] (Table 4).

Table 4. Spectral data on some heterocyclic biphenylene analogs

Compound	NMR chemical shifts (solvent, τ, order as indicated in the structure)	Electronic spectra (λ max, solvent, nm)	Ref.
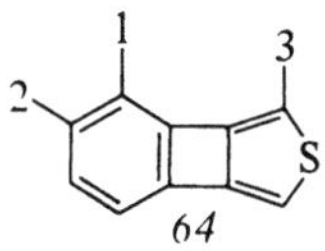 64	$CDCl_3$; 3.14, 3.14, 3.51.	CH_3CH_2OH; 245, 330, 347.	[81]
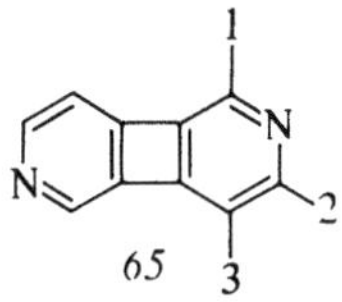 65	$CDCl_3$; 2.0, 1.7, 3.2.	CH_3OH; 241, 283, 297, 312, 325, 338.	[82]
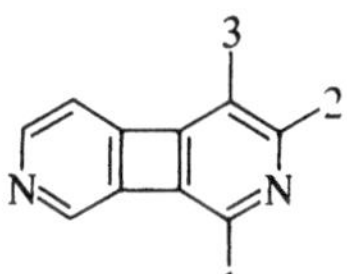 66	$CDCl_3$; 2.10, 1.75, 3.25.	CH_3OH; 234.5, 242, 284, 297.5, 312, 326.5.	[83]
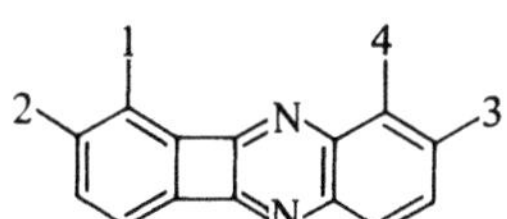 67	CCl_4; 2.50–2.87 (H_{1-3}), 2.43.	CH_3CH_2OH; 232, 237.5, 270, 291.5, 354, 370, 390.	[84, 85]
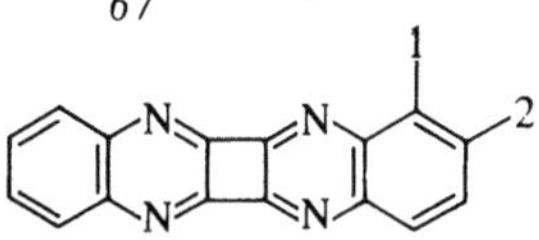 68	$(CH_3)_2SO$; 1.94, 2.14.	CH_3CN; 280, 413, 443.	[86]
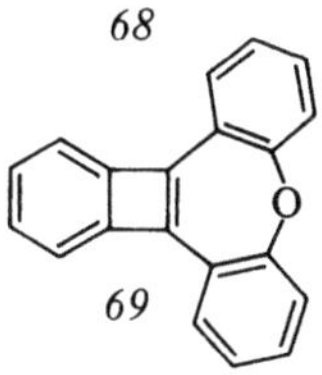 69		THF; 496.	[87]
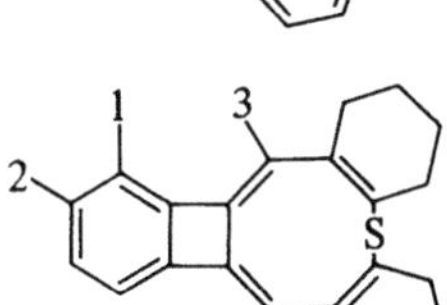 70	CCl_4; 2.91, 2 91, 4.44.	CH_3CH_2OH; 268, 375.	[88]
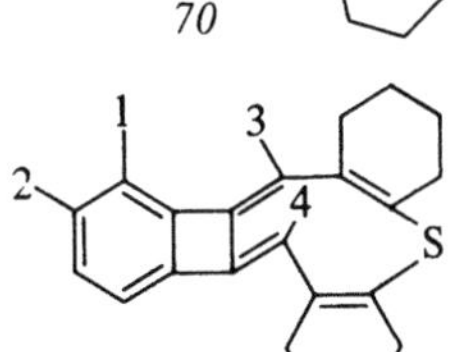 71	CCl_4; 2.92, 2.92, 4.17, 4.38.	CH_3CH_2OH; 314.	[88]

2-Thianorbiphenylene *64*, a heterocyclic equivalent of the norbiphenylene anion *58*, seems to represent a true biphenylene analog according to its physical and chemical properties[81]. In particular, comparison of its NMR with biphenylene *56* and cyclobuta[*c*]thiophene *41* reveals that all the absorptions in *64* have shifted to higher field. Indication that this is due to the induced paramagnetic ringcurrent arising from the four-ring, is obtained on oxidation of *64* to the sulfone *72* which "freezes" the molecule in a dimethylene benzocyclobutene structure. Consequently, a considerable downfield shift of the benzenic absorptions, due to removal of cyclobutadienoid character in the four-ring is observed. The heterocyclic ring protons, however, stay relatively unshifted since they not only experience the loss of para-

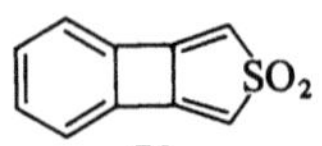

72

tropicity of the central ring, but also the loss of diamagnetism of the five-membered ring, two effects counteracting each other. An attempt to confirm this finding by ^{13}C-NMR spectroscopy revealed that the ringcurrent effects on the ^{13}C chemical shifts, particularly of the annular carbons, are negligible[89].

A dithiabisnorbiphenylene *73* has also been obtained recently[95] and the extent of the paratropic contribution of the cyclobutadienoid ring must await the synthesis of less substituted derivatives. The electronic spectrum of *73* indicates the presence of some interaction between the two thiophene rings.

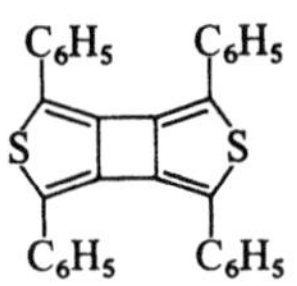

73

The three diazabiphenylenes *65*[82], *66*[83] and *67*[84, 85] seem to exhibit only very moderate upfield shifts when compared with suitable substituted pyridine models[90, 98]. The electronic perturbation of the π-system by several nitrogen atoms might in these systems lead to decreased cyclobutadienoid character. This effect is expected to be even greater in more highly substituted and annelated systems[91] including the tetraaza-analog *68*[86]. The 3-oxahomobiphenylene derivative *69*, however, must have considerable cyclobutadienoid character and its reactivity approaches that of benzocyclobutadiene[87, 92]. Thus, at room temperature dimerization occurs, and the reactive cyclobutene bond can be readily added to dienes and cleaved by oxidizing agents. Unfortunately, no NMR spectrum has been reported.

In contrast to *69*, the all-*cis* and mono-*trans* thionins *70* and *71* are non-planar and have to be viewed as dimethylene benzocyclobutene derivatives of presumably insignificant cyclobutadienoid character[88].

Acknowledgements. I want to express my sincerest gratitude for the constant encouragement and enthusiasm, the unlimited support and the continuous guidance I received from my teachers in chemistry, Dr. P. J. Garratt, Professor F. Sondheimer and Professor R. G. Bergman. I would also like to thank Drs. P. J. Garratt and M. S. Gibson for reading the manuscript.

5. References

[1] Sondheimer, F., Wolovsky, R.: J. Amer. Chem. Soc. *84*, 260 (1972); Sondheimer, F., Calder, I. C., Elix, J. A., Gaoni, Y., Garratt, P. J., Grohmann, K., di Maio, G., Mayer, J., Sargent, M. V., Wolovsky, R.: Special Publication No. 21. The Chemical Society, p. 75. London: 1967; Sondheimer, F.: Pure Appl. Chem. *7*, 363 (1963); Proc. Roy. Soc. *A 297*, 173 (1967); Proc. Robert A. Welch Found. Conf. Chem. Res. *12*, 125 (1968); Accounts Chem. Res. *5, 81* (1972); Haddon, R. C., Haddon, V. R., Jackman, L. M.: Topics in Current Chemistry *16*, 103 (1971).

[2] Kekulé, A.: Bull. Soc. Chim. Paris, [2], *3*, 98 (1865); Annalen *137*, 129 (1866).

[3] Armit, J. W., Robinson, R.: J. Chem. Soc. *127*, 1604 (1925).

[4] Hückel, E.: Z. Physik *70*, 204; *72*, 310 (1931).

[5] Streitwieser, A.: Molecular Orbital Theory for Organic Chemists. New York: Wiley & Sons 1961.

[6] Garratt, P. J., Vollhardt, K. P. C.: Aromatizität. Stuttgart: Georg Thieme Verlag 1973.

[7] Breslow, R.: Chem. Eng. News *43* (26), 90 (1965); Acc. Chem. Res. *6*, 393 (1973).

[8] Dewar, M. J. S.: Advan. Chem. Phys. *8*, 121 (1965).

[9] a) Cava, M. P., Mitchell, M. J.: Cyclobutadiene and Related Compounds. Academic Press: New York (1967);
b) Garratt, P. J., Sargent, M. V.: Advan. Org. Chem. *6*, 1 (1969);
c) Figeys, H. P.: Topics Carbocyc. Chem. *1*, 269 (1969).

[10] a) Maier, G.: Angew. Chem. *86*, 491 (1974); Angew. Chem. internat. Edit. *13*. 425 (1974);
b) Lin, C. Y., Krantz, A.: Chem. Commun. *1972*, 111; Chapman, O. L., McIntosh, C. L., Pacansky, J.: J. Amer. Chem. Soc. *95*, 614 (1973); Chapman, O. L., De La Cruz, D., Roth, R., Pacansky, J.: J. Amer. Chem. Soc. *95*, 1337 (1973); Masamune, S., Suda, M., Ona, H., Leichter, L. M.: Chem. Commun. *1972*, 1268; Maier, G., Hoppe, B.: Tetrahedron Lett. *1973*, 861.

[11] Wittig, G., Weinlich, J.: Chem. Ber. *98*, 471 (1965); Applequist, D. E., Gebauer, P. A., Gwynn, D. E., O'Connor, L. H.: J. Amer. Chem. Soc. *94*, 4272 (1972).

[12] Masamune, S., Nakamura, N., Suda, M., Ona, H.: J. Amer. Chem. Soc. *95*, 8481 (1973); also: Delbaere, L. T. J., James, M. N. G., Nakamura, N., Masamune, S.: J. Amer. Chem. Soc. *97*, 1973 (1975); Brown, R. S., Masamune, S.: Can. J. Chem. *53*, 972 (1975).

[13] Kimling, H., Krebs, A.: Angew. Chem. *84*, 952 (1972); Angew. Chem. internat. Edit. *11*, 932 (1972).

[14] Maier, G., Alzérreca, A.: Angew. Chem. *85*, 1056 (1973); Angew. Chem. internat. Edit. *12*, 1015 (1973); see also: Lauer, G., Müller, C., Schulte, K.-W., Schweig, A., Maier, G., Alzérreca, A.: Angew. Chem. *87*, 194 (1975); Angew. Chem. internat. Edit. *14*, 172 (1975).

[15] Lauer, G., Müller, C., Schulte, K. W., Schweig, A., Krebs, A.: Angew. Chem. *86*, 597 (1974); Angew. Chem. internat. Edit. *13*, 544 (1974).

[16] Dewar, M. J. S.: Angew. Chem. *83*, 859 (1971); Angew. Chem. internat. Edit. *10*, 761 (1971).

[17] Breslow, R., Grubbs, R., Murahashi, S.: J. Amer. Chem. Soc. *92*, 4139 (1970); Breslow, R., Murayama, D. R., Murahashi, S. I., Grubbs, R.: J. Amer. Chem. Soc. *95*, 6688 (1973).

[18] Simmons, H. E., Anastassiou, A. G., in: Cava, M. P., Mitchell, M. J.: Cyclobutadiene and Related Compounds, Academic Press, New York (1967).

[19] Manatt, S. L., Roberts, J. D.: J. Org. Chem. *24*, 1336 (1959).

[20] Gheorghiu, M. D., Hoffmann, R.: Rev. Roum. Chim. *14*, 781 (1969).

[21] Gompper, R., Seybold, G.: Angew. Chem. *80*, 804 (1968); Angew. Chem. internat. Edit. *7*, 824 (1968).

[22] Neuenschwander, M., Niederhauser, A.: Chimia *22*, 491 (1968); Helv. Chim. Acta *53*, 519 (1970).

[23] Hoffmann, R.: Chem. Commun. *1969*, 240; see also Weiss, R., Murrell, J. N.: Tetrahedron *27*, 2877 (1971); Pittman, C. U., Douglas, K. L., Ng, Q. Y., Hunter, W., Pace, D., Kispert, L. D.: J. Org. Chem. *40*, 2121 (1975).

24) Lindner, H. J., von Gross, B.: Angew. Chem. *83*, 489 (1971); Angew. Chem. internat. Edit. *10*, 490 (1971); Chem. Ber. *107*, 598 (1974).
25) a) Gompper, R., Holsboer, F., Schmidt, W., Seybold, G.: J. Amer. Chem. Soc. *95*, 8479 (1973).
b) Gompper, R., Wagner, H.-U.: private communication.
26) Wagner, H.-U.: Angew. Chem. *85*, 920 (1973); Angew. Chem. internat. Edit. *12*, 848 (1973).
27) Seybold, G., Jersak, U., Gompper, R.: Angew. Chem. *85*, 918 (1973); Angew. Chem. internat. Edit. *12*, 847 (1973).
28) Longuet-Higgins, H. C., Orgel, L. E.: J. Chem. Soc. *1956* (1969).
29) Criegee, R., Schröder, G.: Liebigs Ann. Chem. *623*, 1 (1959), Criegee, R., Angew. Chem. *74*, 703 (1962); Angew. Chem. internat. Edit. *1*, 519 (1962).
30) See, Maitlis, P. M.: Advan. Organometal. Chem. *4*, 95 (1966); Efraty, A., J. Organometal. Chem. *57*, 1 (1973).
31) Watts, L., Fitzpatrick, J. D., Pettit, R.: J. Amer. Chem. Soc. *87*, 3253 (1965).
32) Dewar, M. J. S.: Bull. Soc. Chim. France *18*, C79 (1951); Pettit, R., Emerson, G. F., Mahler, J. E.: J. Chem. Ed. *40*, 175 (1963).
33) Sanders, A., Giering, W. P.: J. Amer. Chem. Soc. *96*, 5247 (1974); *97*, 919 (1975).
34) Hess, B. A., Schaad, L. J.: Aust. J. Chem. *25*, 2231 (1972).
35) Blomquist, A. T., Bottomley, C. G.: J. Amer. Chem. Soc. *87*, 86 (1965).
36) Avram, M., Constantinescu, D., Dinulescu, I. G., Constantinescu-Simon, O., Mateescu, G. D., Nenitzescu, C. D.: Rev. Roum. Chim. *15*, 1097 (1970).
37) Avram, M., Constantinescu, D., Dinulescu, I. G., Nenitzescu, C. D.: Tetrahedron Lett. *1969*, 5215.
38) Straub, H.: Angew. Chem. *86*, 412 (1974); Angew. Chem. internat. Edit. *13*, 405 (1974).
39) Ota, K., Murofushi, K., Hoshi, T., Inoue, H.: Tetrahedron Lett. *1974*, 1431.
40) Adger, B. M., Keating, M., Rees, C. W., Storr, R. C.: Chem. Commun. *1973*, 19; Adger, B. M., Rees, C. W., Storr, R. C.: J. C. S. Perkin I *1975*, 45.
41) Emerson, G. F., Watts, L., Pettit, R.: J. Amer. Chem. Soc. *87*, 131 (1965).
42) Victor, R., Ben-Shoshan, R.: Chem. Commun. *1974*, 93.
43) Sanders, A., Magatti, C. V., Giering, W. P.: J. Amer. Chem. Soc. *96*, 1610 (1974).
44) Breslow, R., Washburn, W., Bergman, R. G.: J. Amer. Chem. Soc. *91*, 196 (1969); see also Bauld, N. L., Dahl, C. E., Rim, Y. S.: J. Amer. Chem. Soc. *91*, 2787 (1963).
45) Breslow, R.: Pure and Appl. Chem. *28*, 111 (1971).
46) Breslow, R., Washburn, W.: J. Amer. Chem. Soc. *92*, 427 (1970).
47) Breslow, R.: Acc. Chem. Res. *6*, 393 (1973).
48) Bergman, R. G., Acc. Chem. Res. *6*, 31 (1973).
49) D'Amore, M. B., Bergman, R. G.: J. Amer. Chem. Soc. *91*, 5694 (1969); D'Amore, M. B., Bergman, R. G., Kent, M., Hedaya, E.: Chem. Commun. *1972*, 49; Henry, T. J., Bergman, R. G.: J. Amer. Chem. Soc. *94*, 5103 (1972).
50) Vollhardt, K. P. C., Bergman, R. G.: J. Amer. Chem. Soc. *94*, 8951 (1972).
51) Vollhardt, K. P. C., Bergman, R. G.: J. Amer. Chem. Soc. *95*, 7538 (1973).
52) Bergman, R. G., Vollhardt, K. P. C.: Chem. Commun. *1973*, 214.
53) Garratt, P. J., Nicolaides, D. N.: J. Org. Chem. *39*, 2222 (1974).
54) The assignments were corroborated by the preparation of 1,2-dideuteriofuro[3,4-*c*]cyclobutadiene from the corresponding dideuterated acetylenic precursor; Vollhardt, K. P. C., Bergman, R. G.: unpublished results.
55) Vollhardt, K. P. C.: Synthesis (1975).
56) Garratt, P. J., Vollhardt, K. P. C.: J. Amer. Chem. Soc. *94*, 1022 (1972).
57) For an analogous reaction of (*22*) with TCNE see, Huth, A., Straub, H., Müller, E.: Tetrahedron Lett. *1973*, 2845.
58) Breslow, R., Oda, M., Sugimoto, T.: J. Amer. Chem. Soc. *96*, 1639 (1974).
59) Roberts, B., Wissner, A.: J. Amer. Chem. Soc. *92*, 6382 (1970).
60) Schröder, G., Ramadas, S. R., Nikoloff, P.: Chem. Ber. *105*, 1072 (1972); see also Schröder, G., Röttele, H.: Angew. Chem. *80*, 665 (1968); Angew. Chem. internat. Edit. *7*, 635 (1968).

61) Roberts, B. W., Wissner, A.: J. Amer. Chem. Soc. *94*, 7168 (1972).

62) Fernandez-Alonso, J. I., Domingo, R.: An. Real. Soc. Espan. Fis. Quim. *62B*. 1315 (1966); Hilpern, J. W.: Mol. Phys. *9*, 295 (1965); A. Y. Meyer, Tetrahedron *24*, 6215 (1968).

63) Dewar, M. J. S., Gleicher, G. J.: Tetrahedron *21*, 1817 (1965).

64) Cava, M. P.: Special Publication No. 21, The Chemical Society, London, p. 163 (1967); Cava, M. P., Firouzabadi, H., Krieger, M.: J. Org. Chem. *39*, 480 (1974).

65) Miyamoto, T., Odaira, Y.: Tetrahedron Lett, *1973*, 43; Miyamoto, T., Tanaka, S., Odaira, Y.: J. C. S. Perkin I *1973*, 138.

66) Cava, M. P., Mangold, D.: Tetrahedron Lett. *1964*, 1751.

67) Barton, J. W., in: Nonbenzenoid Aromatics, (ed. J. P. Snyder), Vol. 1, p. 32 New York: Academic Press 1969.

68) Garratt, P. J.: Topics in Nonbenzenoid Aromatic Chemistry *1*, 95 (1973).

69) Lee, H. S.: Chemistry (Taipei) *1962*, 137; *1963*, 22; *1963*, 47; Ali, M. A., Coulson, C. A.: Tetrahedron, *10*, 41 (1960).

70) Yokozeki, A., Wilcox, C. F., Bauer, S. H.: J. Amer. Chem. Soc. *96*, 1026 (1974); Fawcett, J. K., Trotter, J.: Acta Cryst. *20*, 87 (1966).

71) Katritzky, A. R., Reavill, R. E.: Rec. Trav. Chim. *83*, 1230 (1960).

72) Binsch, G., Tamir, I.: J. Amer. Chem. Soc. *91*, 2450 (1969).

73) Farnum, D. G., Atkinson, E. R., Lothrop, W. C.: J. Org. Chem. *26*, 3204 (1961).

74) Bedford, A. F., Carey, J. G., Millar, I. T., Mortimer, C. T., Springall, H. D.: J. Chem. Soc. 3895 (1962).

75) Figeys, H. P.: Chem. Commun. *1967*, 495; Angew. Chem. *80*, 629 (1968); Angew. Chem. internat. Edit. *7*, 642 (1968).

76) Cava, M. P., Narasimhan, K., Zeiger, W., Radonovich, L. J., Glick, M. D.: J. Amer. Chem. Soc. *91*, 2378 (1969).

77) Lombardo, L., Wege, D.: Tetrahedron Lett. *1972*, 4859. For a benzoderivative see, Garratt, P. J., Vollhardt, K. P. C.: Chem. Commun. *1971*, 1143.

78) Baxter, C. S., Garratt, P. J., Vollhardt, K. P. C.: J. Amer. Chem. Soc. *91*, 7783 (1969); see also, Garratt, P. J., Vollhardt, K. P. C., Mitchell, R. H.: J. Chem. Soc. (C) *1970*, 2137.

79) Garratt, P. J., Vollhardt, K. P. C.: Angew. Chem. *83*, 111 (1971); Angew. Chem. internat. Edit. *10*, 125 (1971).

80) Garratt, P. J. Jones, A. J., Vollhardt, K. P. C.: unpublished results.

81) Garratt, P. J. Vollhardt, K. P. C.: Chem. Commun. *1970*, 109; J. Amer. Chem. Soc. *94*, 7087 (1972).

82) Kramer, J., Berry, R. S.: J. Amer. Chem. Soc. *94*, 8336 (1972).

83) MacBride, J. A. H.: Chem. Commun. *1974*, 359.

84) Cava, M. P., Napier, D. R., Pohl, R. J.: J. Amer. Chem. Soc. *85*, 2076 (1963).

85) Vollhardt, K. P. C.: unpublished results.

86) Hünig, S., Pütter, H.: Angew. Chem. *84*, 481 (1972); Angew. Chem. internat. Edit. *11*, 433 (1972).

87) Cava, M. P., Buck, K. T.: J. Amer. Chem. Soc. *95*, 5805 (1973).

88) Garratt, P. J., Holmes, A. B., Sondheimer, F., Vollhardt, K. P. C.: J. Amer. Chem. Soc. *92*, 4492 (1970); J. C. S. Perkin I *1973*, 2253.

89) Jones, A. J., Garratt, P. J., Vollhardt, K. P. C.: Angew. Chem. *85*, 260 (1973); Angew. Chem. internat. Edit. *12*, 241 (1973).

90) Jackman, L. M., Sternhell, S.: Nuclear magnetic resonance spectroscopy in organic chemistry. New York: Pergamon Press 1969.

91) See, for instance, Blatchly, J. M., McOmie, J. F. W., Thatte, S. D.: J. Chem. Soc. *1962*, 5090, McOmie, J. F. W., West, D. E.: J. Chem. Soc. (C) *1970*, 1084.

92) Strictly speaking compound (*69*) is not a true biphenylene analog, since one of the benzene rings of biphenylene has been replaced by a $4n$ rather than $(4n + 2)$ π-electron system.

93) Wilcox, C. F., Uetrecht, J. P., Grantham, G. D., Grohmann, K. G.: J. Amer. Chem. Soc. *97*, 1914 (1975).

94) Irngartinger, H., Rodewald, H.: Angew. Chem. *86*, 783 (1974); Angew. Chem. internat. Edit. *13*, 740 (1974).

95) Garratt, P. J., Neoh, S. B.: J. Org. Chem. *40*, 970 (1975).
96) Footnote 22 in Staley, S. W., Cramer, G. M., Orvedal, A. W.: J. Amer. Chem. Soc. *96*, 7433 (1974).
97) For a thiacyclobutadiene, see: Devdhar, R. S., Gogte, V. N., Tilak, B. D.: Tetrahedron Lett. *1974*, 3911.
98) For monoazabiphenylenes, see: Barton, J. W., Walker, R. B.: Tetrahedron Lett. *1975*, 569.

Received March 27, 1975